101 Careers in Mathematics

Second Edition

Library of Congress Card Number 2002114178

ISBN 0-88385-728-6

Printed in the United States of America

Current Printing (last digit):
10 9 8 7 6 5 4 3 2

101 Careers in Mathematics

Second Edition

Edited by

Andrew Sterrett

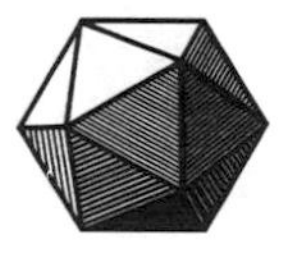

Published and Distributed by

The Mathematical Association of America

CLASSROOM RESOURCE MATERIALS

Classroom Resource Materials is intended to provide supplementary classroom material for students — laboratory exercises, projects, historical information, textbooks with unusual approaches for presenting mathematical ideas, career information, etc.

Published by
The Mathematical Association of America

101 Careers in Mathematics, edited by Andrew Sterrett

Archimedes: What Did He Do Besides Cry Eureka?, Sherman Stein

Calculus Mysteries and Thrillers, R. Grant Woods

Combinatorics: A Problem Oriented Approach, Daniel A. Marcus

Conjecture and Proof, Miklós Laczkovich

A Course in Mathematical Modeling, Douglas Mooney and Randall Swift

Cryptological Mathematics, Robert Edward Lewand

Elementary Mathematical Models, Dan Kalman

Environmental Mathematics in the Classroom, edited by B. A. Fusaro and P. C. Kenshaft

Geometry From Africa: Mathematical and Educational Explorations, Paulus Gerdes

Interdisciplinary Lively Application Projects, edited by Chris Arney

Inverse Problems: Activities for Undergraduates, Charles W. Groetsch
Laboratory Experiences in Group Theory, Ellen Maycock Parker
Learn from the Masters, Frank Swetz, John Fauvel, Otto Bekken, Bengt Johansson, and Victor Katz
Mathematical Modeling in the Environment, Charles Hadlock
Ordinary Differential Equations: A Brief Eclectic Tour, David A. Sánchez
A Primer of Abstract Mathematics, Robert B. Ash
Proofs Without Words, Roger B. Nelsen
Proofs Without Words II, Roger B. Nelsen
A Radical Approach to Real Analysis, David M. Bressoud
She Does Math!, edited by Marla Parker
Solve This: Math Activities for Students and Clubs, James S. Tanton

MAA Service Center
P.O. Box 91112
Washington, DC 20090-1112
1-800-331-1MAA FAX: 1-301-206-9789

Contents

Appendices

Preface to the First Edition

The authors of the essays in this volume describe a wide variety of careers for which a background in the mathematical sciences is useful. They provide more than 101 answers to the question often asked by students, "Why study math?"

These mathematicians are found:

in well-known companies — IBM, AT&T Bell Laboratories, and American Airlines;

in some surprising places — FedEx Corporation, L.L. Bean, and Perdue Farms Incorporated;

in government agencies — Bureau of the Census, Department of Agriculture, and NASA Goddard Space Center;

in the arts — sculpture, music, and television;

in the professions — law and medicine; and

in education — elementary, secondary, college, and university.

Many of these mathematicians started their own companies.

Most of the writers in this volume use the mathematical sciences on a daily basis in their work; others rely on the general problem-solving skills acquired in their mathematics courses as they deal with complex issues. Many mathematicians refer to the importance of communicating with colleagues — working on a team, writing reports, and giving oral presentations. Statistics and computer science, as well as a knowledge of a field where mathematics is applied, frequently are cited as important in one's background.

The degrees earned by these authors range from bachelors to masters to the PhD, in approximately equal numbers. Many with a Bachelor's degree in Mathematics have earned graduate degrees in a related field, often statistics or computer science.

Students should not overlook the articles in the Appendix that are reprinted from the MAA's magazine for students, *Math Horizons*. These articles provide valuable advice on looking for a job and on the expectations of industry.

Andrew Sterrett
Emeritus Professor of Mathematical Sciences, Denison University
Visiting Mathematician, Mathematical Association of America

Preface to the Second Edition

After the publication of 100 career profiles in the first edition of *101 Careers in Mathematics*, thirty-one additional profiles were obtained and are included in this edition. The new profiles also represent a wide variety of academic backgrounds and jobs and, in order that one might better see what advanced degrees (if any) undergraduate mathematics majors go on to earn, an appendix, Degrees Earned by Mathematics Majors, is included in this edition. Perhaps the most surprising observation to be made from this listing is found in the large number of mathematics majors who have advanced degrees in other disciplines. Mathematics majors are seen to go on to use their backgrounds in mathematics in conjunction with almost every other discipline, telecommunications, ecology, law, and economics to name a few.

Of course many individuals with a mathematics major or a keen interest in mathematics stray quite far from mathematics. It should not be surprising that an agile, seven-foot tall math major and basketball player would find his way to a professional basketball team. A 1993 list of famous nonmathematicians, compiled by Steven G. Buyske, is found in an appendix to this volume. Up-to-date sources for famous nonmathematicians on the web include:

www.geocities.com/SoHo/Square/7921/fammaj.html (with photos)

www.stat.rutgers.edu/~buyske/nonmath.html

www.stcloudstate.edu/~dbuske/famousnonmathematicians.html

In addition to new profiles, this edition contains "updates" (indicated by an * in the Table of Contents) written by some of the original authors. These updates illustrate that one's first job is not always the last one and that major changes in one's career path sometimes occur.

Although this volume is intended primarily for students, faculty members might benefit from it as well. In his review of *101 Careers in Mathematics* which appeared in the *American Mathematical Monthly* (104, 1997, pp. 579–482), J. Kevin Colligan wrote, "Get a copy of *101 Careers in Mathematics* to see what comes out the far end of the mathematics education pipeline. It's thought provoking."

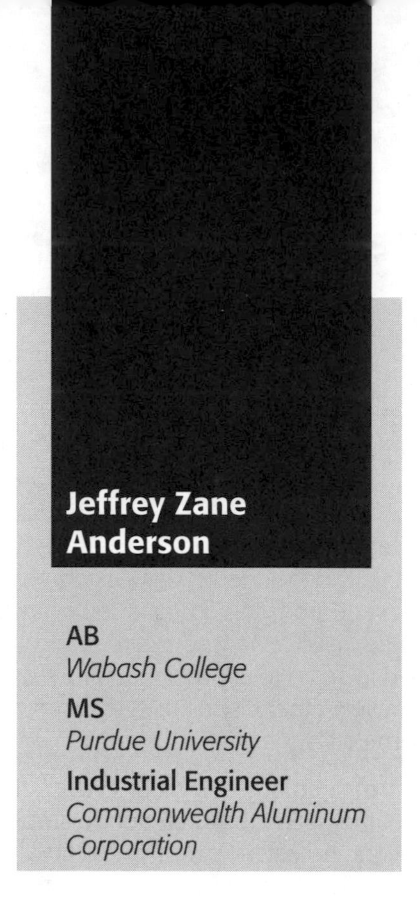

Jeffrey Zane Anderson

AB
Wabash College

MS
Purdue University

Industrial Engineer
Commonwealth Aluminum Corporation

Although I had an aptitude for mathematics during my elementary and secondary education, my interest in mathematics did not begin until I was a student at Wabash College. As a college student, I was attracted to mathematics by the logic and discipline required to successfully complete complex mathematics problems. During a study abroad program in Aberdeen, Scotland, my interest in mathematics was enhanced by an introduction to the applications of linear algebra and optimization. This interest, after graduation from Wabash, led me to Purdue University, where I completed a MS degree from the School of Industrial Engineering. At Purdue, I specialized in operations research, which is broadly defined as the application of mathematical models to business and engineering problems. My study of mathematics and related subjects has uniquely qualified me for my industrial engineering career.

As an industrial engineer for Commonwealth Aluminum Corporation, a rolled aluminum sheet manufacturer, I draw daily upon the skills acquired from my mathematics education. At Commonwealth, my job duties require that I unite the concepts of mathematics, operations research, and industrial engineering to conduct computer simulation and data analysis projects.

Capacity determination is one application of computer simulation that Commonwealth utilizes. For example, working with other industrial engineers, we developed a comprehensive simulation model of our casting facilities. This model enables us to test the effects of process improvements and facility modifications before any capital expenditure is made. Construction of such a model requires the ability to translate 'real world' processes into a simulation program. Then the appropriate selection of the statistical distributions to apply in this program must be made. By studying mathematics, I have been provided with the necessary logic and analytical skills to complete these tasks.

Yet another way that I apply mathematics in my industrial engineering career is through data analysis. Effective data analysis requires competency in choosing correct analysis variables and presentation of the results in a manner which non-technical management can readily understand. I regularly use commercially produced statistical analysis software packages and spreadsheets to complete the required analysis. By understanding the rationale behind the algorithms that these software packages apply, I am able to more effectively complete the analysis and clearly communicate the results.

Being adaptive and well versed in computer applications, of which mathematics is an integral component, is instrumental in the field of industrial engineering. Without a firm understanding of mathematics, meeting this criteria would be unattainable. By studying mathematics, I have gained the confidence in my quantitative and analytical abilities to be an effective industrial engineer.

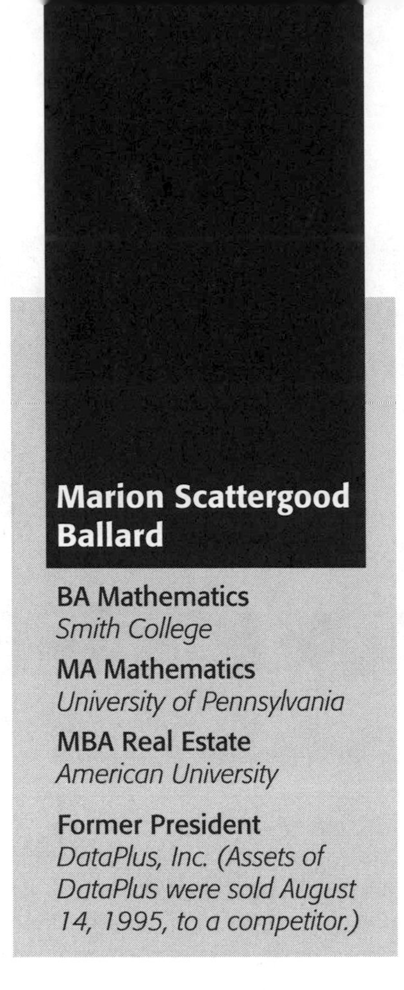

Marion Scattergood Ballard

BA Mathematics
Smith College

MA Mathematics
University of Pennsylvania

MBA Real Estate
American University

Former President
DataPlus, Inc. (Assets of DataPlus were sold August 14, 1995, to a competitor.)

It is 6:00 am. My husband wakes me to join him for our daily health walk — one hour down the bike path and up the rail trail. We greet the regulars and wonder who each new person might be. I am at my desk by 8:30, having dropped my husband off at the metro. The phone begins ringing. Barbara S. wants to know how to print a list of donors who have given $50 once and only once to annual giving. No one from the support team is in, so I take the call. Proofs of the newsletter are back and need reading.

[After graduate school, I did some exciting work at Univac, programming a Univac II to calculate coverage of the earth by the manned orbital laboratory.]

Martha C. would like a quote on adding security to her prospect management system. The next call is an inquiry from a prospect researcher who saw our full-page ad in the Chronicle of Philanthropy. It is hard to be enthusiastic about the prospect of a new client when you anticipate the sale of your business.

[I went into business for myself in 1980, writing customized programs and selling and installing computers and networks. My best customers were non-

profit organizations who needed computer programs to track donors and their gifts. Last summer there were 10 of us. A couple of weeks ago we got an unsolicited offer for a buy-out by our biggest competitor. Too good to turn down.]

My lawyer is on the phone. What do I think of his memo to the lawyer of the acquiring company? We schedule a five-way conversation to discuss the final points in the sales agreement.

[A couple of months ago I bought a trip through the Northwest passage on a Russian icebreaker — on a whim at a charity auction.]

I need to buy film and a polarizing lens for my icebreaker trip, which starts in two days. I miss a call from the lawyer who is trying to finish up my will before I leave. I don't get to the camera store before it closes — the caterer for my son's wedding called about the menu for the rehearsal dinner.

[What am I going to do when I return from this once-in-a-lifetime trip? I have just rotated off the Board of SSFS, a growing independent school where I was Chairman of the Board. I've had my business for 15 years. It is time for a change. I need it and I want it. Computers again? Maybe Lotus Notes. Maybe DOS to WINDOWS conversions. Maybe affordable housing, a long-time interest that led to a concentration in real estate for my MBA.]

I am a typical mathematician, doing work typical for people with analytical minds, even though I cannot say this was a typical day.

P. Darcy Barnett

BA Mathematics
Dunbarton College
Computer Scientist
National Institute of Standards and Technology

During the last several years my work has involved system administration of Sun workstations. In general my job is to enhance and supportdistributed computing in the Information Technology Laboratory. I have designed and configured software for various purposes and for various platforms, and helped others do the same. Currently, I am looking forward to embarking on a new phase of distributed computing, namely, the design of and the implementation of the "Distributed Computing Environment" and "Distributed Management Environment" at our site. Other exciting projects involve enterprise-wide email and calendaring services.

My first job after college graduation was with Bellcomm, Inc., a subsidiary of AT&T, which performed systems engineering for the Apollo space program. At that time, computer science was a burgeoning area. In fact, my college did not offer any computer science courses per se: the closest thing with direct applicability was numerical analysis. So I learned Fortran on-the-job and began to do scientific applications programming. The projects included a space craft trajectory generation model, an autopilot for the lunar excursion module, and a spacecraft weights and sensitivities simulation. Physicists and mathematicians with more experience and advanced education did the modeling design, whereas others and I did the programming and implementation. Our results were compared with NASA's and another independent systems engineering group's for consistency and efficiency. Of course, the success of the Apollo 11 moon landing was the culmination of the efforts of many people.

After a year's leave of absence to have a child, I returned briefly to Bellcomm but decided that I did not want to work full time and I found a part-time job (mathematician) as a programmer at the National Bureau of Standards. This was great; it was doable because the programming tasks could be done without having to job-share as is the case in other situations. As my family and life changed, so did my tour of duty at NBS.

During the course of my career, I have consistently taken classes and courses to enrich my knowledge of job-related computer topics. Participating in a degree-granting program is almost essential for increased upward mobility in one's career. Having a doctorate does make a difference in the research and development arena. However, for me the trade-off of spending more time with my children outweighed the benefits of a PhD and career advancement.

It was my mathematics education that prepared me for the rigors of designing and implementing the programs for large-scale simulations and solutions of systems of PDEs and for the logical thinking and planning necessary to debug and develop the components (internals) of computer operating systems. However, it is of great importance to develop your communication skills, written and verbal. No matter how great your ideas are, if you cannot convey them to the audience, it will be difficult to get the audience on your side. It will be very frustrating to see your ideas ignored and the lesser ideas of a more articulate person gain acceptance.

Carl M. Beaird

BS/MS Applied Mathematics
BS/MS Physics
California State University,
Los Angeles

Mathematician
Rockwell International
Corporation

After an honorable discharge from the US Navy, employment in electrical prototype and testing for the defense industry beckoned. After marriage and children, I decided to reenter college.

Having a love for engineering, electrical to be precise, I applied to California State University, Los Angeles, choosing a major in mathematics and physics. Immediately after receiving my bachelor's degree, I entered graduate school to pursue a Master of Science in applied mathematics with an emphasis on numerical analysis and physics.

The undergraduate courses of study were broad and diverse: problem-solving techniques, logic, mathematical philosophy and approaches; solutions in physics with several approaches and techniques; and computer philosophy and languages. These courses, coupled with general-education required courses, made me a more well-rounded person.

Upon completion of these requirements, I earned a Master of Science degree in applied mathematics. After an arduous job search, I landed at Rockwell's Space Systems Division.

My first assignment was a dynamic one: researching an approach and solution for a frontal area algorithm for the Space Station. Part of education is realizing the absence of depth and expertise needed in certain situations. In this case, help from someone else was required, and my research led to the hiring of an expert in programming techniques and systems. His solution was published nationally.

Some of my other projects involved writing algorithms that graphically detailed the automation of installed and uninstalled Shuttle tiles. I also contributed to the development of a critical component tool for the Space Shuttle. I am currently working on an advanced tool design for space objects using innovative approaches with algorithms.

Diversity and engineering are the two aspects I enjoy most about my career. Without my broad and diverse technical background, the opportunity for adventure would not have unveiled itself. Education in mathematics, the sciences, and related fields offer a world of promise. For me, my career is a "magnificent obsession" fulfilled.

Cherryl Beeman

BS Mathematics
University of Lowell

Programmer/Analyst
Lockheed Martin Defense Systems

Electronic Commerce Specialist
The Kodiak Group

I started college as a computer engineering major at a small private college. Sophomore year I switched colleges (to U. Lowell) and majors (to math). Both decisions were wise. During my senior year, I was interviewed on campus by GE, and hired by the manager of a systems engineering group at Defense Systems in Pittsfield. The manager was specifically looking for someone with a broad mathematical background to do programming and analysis work on projects for requesting engineers. I loved turning abstractions of physical systems and situations (i.e., mathematical models) into efficient and well-designed computer programs. I was immediately set to work learning positional astronomy and writing programs to select the stars to be observed during field testing of our guidance system and sending the necessary positional data to the test site. After that exciting initial assignment, I worked on many different projects, mostly involving the design, revision, and maintenance of large Monte Carlo simulations of various guidance and artillery subsystems as well as the associated data analysis.

A secondary portion of my job was to keep the engineers up-to-date on new computer systems and software through informal training programs and consultation. I briefly considered pursuing a Master's Degree in Systems Engineering and even took a few courses, but decided that I preferred the mathematical aspects of my job and still did not want to become an engineer. During my time

with GE and Lockheed Martin, I often took advantage of the in-plant training program in order to keep current with the rapidly changing Information Technology field. In this way, I obtained a familiarity with new or different operating systems (including UNIX), languages (ADA, C), scientific studies (underwater acoustical modeling) and defense systems. I also spent some time in self-study, and thus learned the MS Office Suite and Visual Basic. The learning process does not stop once the degree is in hand!!

I finally tired of the upheaval and uncertainty of the defense downsizing of the early 90's, and left Lockheed Martin after 11 years. I accepted a position with an Electronic Commerce (EC) consulting company called The Kodiak Group. The environment is quite different from my former job, as my new company is small (but growing: we will be opening a third office in Denver), privately owned and operated by four equal partners. The technical aspects of my job are quite similar: writing applications programs and scripts, developing and managing databases, establishing remote access hookup to client systems, and administering a Lotus Notes network.

I was fortunate to find not just one, but two excellent jobs quickly and easily. When the time came for a change, I was able to take the skill set started in college and developed during my years in a scientific environment with a multinational company and apply them to helping a small company grow and succeed in a business oriented environment. My mathematical training allowed me to adapt very quickly and to be productive almost immediately.

Teaching in a private school has given me a way to spend all of my work time with young people as they learn mathematics. I am grateful to those mathematicians and educators who write texts and articles, create manipulative aids and demonstrations, design and improve scientific and graphing calculators, or provide computer software from which I select what my students need. I like being the person who works with each student, explaining, listening, suggesting and encouraging that individual learner.

In our school (4–12; 600 students), I usually teach four upper-school courses. One of these is a linear algebra and multivariable calculus course, for students from St. Albans and several other schools in the District of Columbia, who have already completed a first year calculus course. All of my classes are small, so it is easy to plan lessons around computer exercises, use of manipulatives, cooperative group work, open discussion or formal lecture. No single format needs to dominate.

I am the planner, choosing my own texts and course materials, inventing and adapting as necessary. I am the lecturer, using prepared notes and enjoying the challenge of the need for impromptu response to student questions. I am the

facilitator, as students work on computer labs or in small groups seeking solutions to very challenging problems. I am the teaching assistant, available for individual dialogue or help during the student's free or study periods as well as during class time. I coach upper- and middle-school math teams each year and groups involved in other academic competitions. I write comments, evaluations and college recommendations. I lead middle school outdoor explorations of the area and frequently participate in student-faculty recreations like playing bridge and going bowling. I work in cooperation with the other teachers, the administration, students and parents as we plan for the total community which is our school. I work formally and informally with the mathematics teachers in other private schools in the area as we share ideas and information. I am an active member of the National Council of Teachers of Mathematics, giving workshops at national and regional meetings, and writing for the journal *Mathematics Teaching in the Middle School*.

My educational background includes an undergraduate major in mathematics from a small liberal arts college and a PhD in mathematics education; I was a participant in the 1992 Woodrow Wilson Mathematics Institute at Princeton University. The content of the courses I teach ranges from elementary algebra to multivariable calculus, and, as the interests of my students lead us to explore a variety of topics in theoretical and applied mathematics, I continue to draw on all that I have studied, and I continue to be a real student of mathematics myself. Every day brings a new challenge!

Jonathan G. Blattmachr

BA Mathematics
Bucknell University
JD (law)
Columbia University School of Law
Partner
Milbank, Tweed, Hadley & McCloy
New York and Los Angeles

Because I had long intended to be a mathematics professor, I gave little thought during high school and college to the practical use of mathematics and mathematical skills in endeavors outside of obvious ones such as engineering. Toward the end of my college days, I developed a great interest in economics. I did very well in economics-related subjects as a result of understanding mathematical principles. A significant portion of economic study revolves around interpreting and analyzing the interrelationships among factors which drive the economics of a particular organization, an industry, or even a country. The interpretation and analysis are tied primarily to mathematical principles. As much as I enjoyed economics, I ultimately decided to go to law school. I never considered that my study of mathematics would help me in law school. I was wrong.

Although I had no background in the law (I had not even taken one course in political science), I did well at one of the nation's best law schools. I attribute much of my academic success at the Columbia University School of Law to having learned, through the study of mathematics, and in particular theorems, how to analyze complicated principles. Comprehending certain laws is as challenging as understanding some of the most complicated mathematical theories you will encounter.

Now I practice law as a partner in one of the country's largest law firms. My work is concentrated in the areas of taxation and estate planning. I feel I have done well in my job, and I attribute much of that success to my facility with numbers

and mathematical theory. Understanding a legal principle is important, but being able to apply it to produce a better result for a client is even more important and is often regarded as the hallmark of an outstanding lawyer. As the tax laws become more restrictive, understanding the full consequences of actions clients undertake is more important than ever.

Many lawyers, because of their inability to understand complicated theoretical concepts, are often bewildered when trying to foresee what the full impact of implementing certain actions will be. I have found that those who have studied mathematics can approach and master both the legal principles and their effect in a way which most others cannot. One of the younger women lawyers with whom I work has a strong mathematical background and has developed into an outstanding young lawyer. It is reinforced to me daily that one of the reasons she practices law so well is her strong facility for concepts —a facility which was developed in her study of mathematics.

In addition to practicing law, I am involved with the development of a computer program which will assist other lawyers in estate planning for their clients. This program will not only allow lawyers to determine quickly the tax effects of plans and how tax principles may suggest a certain course of action for their clients, but also will allow attorneys to prepare documents (such as wills) in a very efficient and timely manner. The colleague with whom I have developed the program also has a strong background in math. We were able to prepare this program only because of our familiarity with mathematical concepts.

Ron Bousquet

BA Mathematics/Computer Science
Potsdam College
MA Mathematics
Potsdam College
R&D Project Manager
Hewlett-Packard Company

I graduated from Potsdam College in May 1989 with both undergraduate and graduate degrees. I immediately began working for Hewlett-Packard as a software development engineer.

Working in software development, it was easy to see how my education in computer science would apply to my daily work. What wasn't quite so obvious was how my mathematics background would contribute. Within my first six months, it became apparent that what I learned in the realm of mathematics might even be more important than what I learned in computer science.

My first major project was to assist in the extension of a basic information retrieval system. Much of this work involved investigation of current research on advanced information retrieval techniques. This meant that I would read a series of articles submitted to technical journals and combine concepts from various articles into our own information retrieval system. I found my mathematics training very useful in understanding the technical details of many of the articles. Also, many articles contained algorithms which were often incomplete or inaccurate; having the mathematical knowledge to notice these inaccuracies proved invaluable.

But what I found most useful about my mathematics training was not simply the facts I learned. Rather, it was that it taught me how to learn. I was able, with little or no supervision, to pick up a journal, or a textbook, or a manual, and get the information I needed to be productive. In fact, this skill is especially useful in the business world where there is really no equivalent to the "teacher" who seems to invariably know all of the answers.

I am still working at Hewlett-Packard, but now as an R&D Project Manager. Instead of actually developing software solutions, I now manage a group of cross-functional engineers who do that job. Again, I have found my educational background an invaluable asset in preparing me for technical management. My computer science education and experience has given me a strong technical foundation while my mathematics background has provided an analytical framework in which to make informed, thoughtful decisions.

Margaret L. Brandeau

BS Mathematics
MIT

MS Operations Research
MIT

PhD Engineering-Economic Systems
Stanford University

Associate Professor of Industrial Engineering
Stanford University

When in college, I was drawn to mathematics by its elegance and simplicity. I was also intrigued by the many ways mathematics can be used to solve complicated and important real-world problems.

After I graduated from college, I worked for a small consulting company. One of my first projects was to implement a mathematical model to help the city of Boston better deploy its ambulances. We considered different home locations for the ambulances (the locations the ambulances return to when idle). For each combination of home locations, the model used a queuing analysis to determine the average time until an ambulance could respond to calls from different parts of the city, the workload of each ambulance, and other performance measures. This information helped city planners determine a good set of home locations for the ambulances.

After several years as a consultant, I went back to school for a PhD. I studied in a program called "Engineering-Economic Systems" that applies mathematical and economic analysis to business, government, and social systems.

Photograph by Edward W. Souza, Stanford University News and Publications Service

Since 1985 I have been a Professor of Industrial Engineering at Stanford University. During that time, I have worked on a variety of research projects, all of which apply mathematical models to help make better decisions for complex problems.

One problem involved determining how to set up machines that insert components onto printed circuit boards. In a typical application, hundreds of boards are produced from thousands of components. The question is to determine which machines should insert which components so as to minimize the total insertion cost. The problem is complicated by different board setup and insertion costs on each machine, different numbers of each type of board that must be produced, capacity limitations of the machines, and other factors. Working with another researcher, I developed an algorithm to solve this problem. We were recently awarded a patent for our algorithm.

I am currently working on AIDS policy analysis. We use a mathematical model of the epidemic (based on a set of simultaneous nonlinear differential equations) to consider the effects of behavioral changes brought about by AIDS interventions. For example, if 10 percent of high-risk women are screened for HIV, the virus that causes AIDS, and those found to be infected reduce their risky behaviors by a given amount, our model can calculate the number of infections that would be averted in the population. This information is combined with an economic model to determine the likely costs and benefits of different interventions aimed at halting the spread of HIV.

Tiffany H. Brennan

BS Applied Mathematics
Appalachian State University

Information Systems Consultant

I earned a BS degree in applied mathematics with a minor in computer science. Just out of school I began a career as an industrial engineer with a well-known textile manufacturing firm. Before my initial interview, I had no idea what might be expected of an industrial engineer. However, since the interviewer had asked specifically for mathematics majors, I decided that the interview would be a good experience. Little did I know that I would build a career in that field for the next several years.

As an industrial engineer, I monitored labor costs in manufacturing through time-study analysis, developed cost standards through work studies, worked on methods to increase productivity, and designed plant layouts on the AUTOCAD (computer-aided design) software program.

From textiles, I went to furniture manufacturing as an industrial engineer. While with this major furniture manufacturer, I was given the opportunity to transfer into the Systems Development department. In this new department as a system specialist, I programmed in COBOL (developing and modifying batch programs). I also developed real-time applications in MODEL 204 Database. Systems development enabled me once again to communicate with both the manufacturing and administrative ends of the business, as engineering had done before.

Personal circumstances soon required that I move to Tampa, Florida. Although I was told repeatedly that Florida was a "service industry state," not a "manufacturing state," and that my background in mathematics and manufacturing wouldn't be to my advantage, I was fortunate to obtain a job with a consulting firm within a month of my move to Tampa.

I was with this firm for $2\frac{1}{2}$ years as an Information Systems Consultant. As a consultant, I billed out to companies for my services on a contract basis. My first and only client was a large major international bank. My responsibilities here included coordinating a portion of a large project to integrate several different software systems into one for multiple products and currencies, monitoring test problem reports for audit, reconciling current versus old system reports, modifying COBOL programs, and helping to convert old systems to new ones.

Since leaving the bank, I have become a homemaker raising three children. I have not returned to the outside work force yet.

If nothing else, my story illustrates the wide range of careers that an applied mathematics degree offers. With each career change, I was able to adapt to a new environment and relate past experiences to current situations. As a result, I have made myself more marketable and have gained a vast knowledge of several industries — even a service industry!

Judith R. Brown

BA Mathematics
MS Mathematics and Education
University of Iowa

Manager
Advanced Research Computing Services
The University of Iowa

Scientific visualization is a rapidly growing field which combines several disciplines—computational science, computer science, cognitive science, computer graphics systems, and the visual arts—in a common search for new scientific insight. This goal is achieved by using computer graphics tools and techniques to examine the enormous amount of data produced by modern scientific simulations or experiments. "Visualization" has become a buzzword, frequently misused. The important concept is that the information and insight gained from the use of computer graphics are important, not the graphics themselves.

My job is to consult with researchers to determine how their data might be visualized and to help bring their data into the available visualization software. What do I like about my job? My job is so exciting to me that I can hardly wait to get to work in the morning. Each project is a different challenge, and, although we can do some very good work on low-level equipment such as Macintoshes, more advanced workstations are now affordable. Working in a university environment offers both drawbacks and advantages. The drawbacks are in the areas of equipment acquisition and salary. Universities have less money than major industries, which means that major equipment acquisitions are more difficult to obtain and take longer, and university salaries are traditionally 10% to 50% lower

than those in industry. On the other hand, flexible working hours, more vacation, and creative freedom in your job are common in the university environment.

My background is a mix of formal and informal training. My formal education includes a BA in mathematics and education, with a curriculum heavy in the sciences, and an MS in theoretical mathematics. The first company I worked for trained me to program computers because this was in 1964 and there was not yet a computer science program in my university. I have also worked as a consultant on computer graphics hardware and software, across disciplines, for 10 years. My arts background is informal. I have taken pottery classes, have been an art museum docent, and have been active in several arts organizations on campus.

Today, however, there are formal computer graphics courses one can and should take in most colleges and universities, and there are beginning to be programs in scientific visualization. If you want to work as a visualization specialist, you should have a computer science background which includes computer graphics and user interface training. You need strong mathematical skills, especially in linear algebra. You also need visual training from the arts, especially in color theory and drawing. A background with a variety of sciences is also important since the projects range across many scientific and nonscientific disciplines.

Computer graphics professions tend to be multidisciplinary, and scientific visualization is no exception. This is an area where mathematics, science and art coalesce. It is also an area where communications skills—visual, oral, and written—are extremely important.

Today, there is an incredible demand and desire in the business world as well as our personal lives to possess the ability to communicate no matter where we are. As a member of the engineering staff at Westinghouse Wireless Solutions Company, I have the exciting and always challenging opportunity to help create groundbreaking wireless communications products. Our latest product, a mobile satellite terminal, provides mobile voice, fax, and data services from anywhere in North America to anywhere in the world. This means people traveling in areas uncovered by cellular networks can now be in touch with civilization by land, sea or air!

At Wireless Solutions, the technical staff works as an "integrated product development team" which includes software, systems, and hardware engineers. We work in a close knit environment to promote design cohesiveness as well as the opportunity for learning across engineering disciplines. As part of this team, my main role has been to design, develop, and test the front-end software of the mobile terminal transceiver. I implement digital signal processing algorithms on specialized microprocessor chips called DSPs. Digital signal processing allows a physical RF (radio frequency) signal to be sampled (quantized into digital words)

and mathematically processed in real time by a computer to recover bits of information. Digital signal processing and communications theory are founded in mathematics such as Fourier analysis, linear algebra, complex variables, and stochastic processes. I enjoy this niche of software development since it combines principles of math, computer science, and electrical engineering.

My undergraduate curriculum was in mathematics with a concentration in applied math, including credits in computer science and physics. After working for a year as a software engineer in Westinghouse Electronic Systems Group, I returned to school at night to obtain a masters degree in electrical engineering with a concentration in digital signal processing. Upon reflection on my education, I feel my training in mathematics provided me with the invaluable ability to apply logic, reason, and careful quantitative, as well as qualitative, analysis to my work. These thought processes along with good written and oral communications skills are desirable and applicable to almost any field. To prepare for a career in digital communications, I recommend a mix of math, computer science, and electrical engineering at the undergraduate level. Remember—a calculating mind is a good mind!

Cora L. Carmody

BA and MA Mathematics
Johns Hopkins University

MS Computer Science
Fairleigh Dickinson University

PhD Electrical Engineering, in process
University of Houston

Manager of Systems Engineering
PRC, Inc.

During my senior year at Johns Hopkins, a professional services company named PRC came to campus to interview students graduating in mathematics. Although at the time I didn't know the first thing about computers or software, the work their managers described sounded interesting, and I joined them upon graduation. That was in 1978, and not a single year has passed that I haven't felt that that was one of my best decisions ever.

My first project with PRC was a digital imagery exploitation system. My first programming assignment was to develop the image measurement capability. This called for menu-driven software to compute areas and volumes of different image features selected by photointerpreters using the system. It was interesting to apply mathematics in this way, and I discovered that programming was great fun. As I progressed as a programmer, I took on other assignments such as a graphics subsystem, coordinating system translations, and character message translations.

Throughout my years with PRC, I have found a tremendous variety of work, all challenging, and very enjoyable. I spent one year with the research and development group studying artificial intelligence and researching methods of reasoning with uncertainty (which vary from probabilistic methods to fuzzy logic). I

spent another year in the systems integration group, working on making hardware and software work together to meet customer requirements.

My current job is actually several jobs in one. I am the Manager of Systems Engineering on a project supporting the Space Station Freedom. In addition to managing a group of PRC systems engineers, I have been working for over a year with the primary developers of the flight software on the process that will be used to develop, integrate and install the flight software on-orbit. I am also working on a small R&D project, which is studying the applicability of that process to other types of software development.

The last of my current jobs is that of a coordinator and trainer in PRC's quality improvement initiative. This program is our version of TQM, and is one of many corporate programs in America which are aimed at regaining our country's competitive edge and reputation for quality products and quality services. PRC's program is based on four very simple principles—customer satisfaction, respect for people, continuous improvement, and management by fact. The last principle, Management by Fact, is where, once again, having a mathematical background is an asset. This principle aims at making data collection and analysis, using various statistical techniques, a part of every employee's arsenal of decision-making aids.

Over the years I have found that a background in math has given me good problem-solving techniques as well as some specific kinds of knowledge needed in the study of computer science. And now, as I embark on a formal study of electrical engineering, again I find that having a math background is an advantage.

Jack Cassidy

BA, Mathematics
Cornell University

Engineer Scientist, Firmware
Hewlett-Packard

I work as a firmware engineeer for Hewlett-Packard. My group develops the algorithms that control All In One devices—printers that scan, copy, and fax. When people ask me what I do all day, I tell them I solve puzzles, little puzzles and big puzzles. At any time, I'm working on several different things with time scales ranging from a few minutes (why did my code fail to build) to several months (develop a paper-moving algorithm for 20% faster printing). Some of the puzzles involve other people (the power supply is overloaded), while others involve just figuring things out by myself (how can I write a perl script to automate a process). The great thing is these puzzles have answers, and it's satisfying to discover them.

Many people study engineering to get a job like mine, but I'm happy I majored in math. I believe my liberal arts major gave me a freer, fuller educational experience. More importantly, mathematics qualified me not just for one job, but for a whole range of possibilities.

I tried several other careers. After college, I spent a year playing poker in California card rooms. I thought it was going to be pure fun, but I found spending my days with card players unpleasant. Many of them only talked about what bad luck they had, and they talked a lot.

I spent some years doing creative writing in an academic setting. That was interesting, and I met wonderful people. But when I took an elective course on interactive computing, it was more interesting than writing.

I became an entrepreneur, developing utility software for the Apple II with a company called Beagle Bros. I learned a lot about computers and had friends who got rich, but most of my work was done alone at my home computer. When I obtained a regular engineering job, it felt like a holiday. I was suddenly surrounded by bright people who enjoyed their work, and I didn't have to feel guilty about not working weekends.

I moved into management for a few years. It gave me a feeling of knowing what was going on in the company and being an Important Person. It was exciting and challenging, but then I took a six month sabbatical. I ended up using the time to write a mathematical paper on the theory of poker (Nov. *98 American Mathematical Monthly*). I enjoyed it so much that I had to get back to solving puzzles on a daily basis. I transferred back to engineering, and have enjoyed the change.

Shane Chalke

BS Mathematics
Worcester Polytechnic Institute

Actuary/Business Owner
Chalke, Inc.

Even though my colleagues advised against it, I opened my own consulting business nine years ago. Chalke Incorporated now has 35 employees, is the insurance industry's largest supplier of asset/liability analysis software, and provides consulting services to hundreds of insurers. I have found my niche developing new methods of analysis, bringing fundamentals from both economics and mathematics to bear on attacking business problems.

The path that led to my current position began with my deep interest in mathematics. I have been driven toward mathematics for as long as I can remember. By the time I was 13 or 14 years old, I knew that math was my forte. In junior high school I participated in an experimental math program that spurred my interest further.

During high school, I was a member of the math team and participated in a "5th year" math program. In college, I maintained my interest in mathematics by concentrating my studies in the area of probability and statistics. At Worcester Polytechnic Institute I learned about actuarial science. I took courses in life contingencies and theory of interest and developed an interest in insurance. I worked with WPI's actuarial science professor on an independent basis and designed and developed a life insurance product.

After college, I worked at State Mutual as an actuarial student for several years, where I gained a broad exposure to all aspects of life insurance. My second professional experience took me to Massachusetts Mutual, where I designed

and developed a series of "non-traditional" life insurance products. I researched, modeled, and completed financial analysis for many new-wave products of my day. My last stepping stone before starting my own firm was a position at Transamerica Occidental in Los Angeles. While at Transamerica, I finished my actuarial exams and became a Fellow of the Society of Actuaries. For two years I managed Transamerica's Actuary Research Department, where I explored new methods of product development and financial analysis.

It was during my tenure at Transamerica that one of my associates, Michael Davlin, gave me the most valuable gift of my career. He introduced me to the rich field of Austrian economics and showed me the importance of integrating mathematical modeling techniques with fundamental economic principles. It is the integration of these two disciplines that forms the basis of my success today.

After I attained Fellowship status, I left Transamerica and started my own business. I was discouraged by nearly everyone I spoke with ... either I was too young, the timing was wrong, the West Coast wasn't the right place, etcetera. Since I didn't receive much outside support for my decision, I had to depend on myself.

One of the positive aspects of an actuarial career is that personal development is determined more by performance than by external factors. I had a great deal of control over my advancement. I took the actuarial exams at a fast pace while I was working in various actuarial positions. In the end, I opted for the ultimate level of personal responsibility and control over my future.

Judith E. Chapman

BS Mathematics
Cedar Crest College

MA Applied Mathematics
University of Maine

Senior Software Engineer
Harris Scientific Calculations

During my career as a mathematician I have studied glaciers, worked as a teacher, studied optical analysis, and learned about "stealthy" planes. Now I am going to learn about electrical engineering and CAD technology. To me, that is the best thing about having a degree in mathematics – my career can follow any path I want it to.

After completing my Bachelor's degree I decided to pursue a Master's in applied mathematics at the University of Maine. Once at UMaine, I had to decide whether to do my MA research jointly with a fluid mechanics, chemistry, image/technology, or glaciology research group. I chose glaciology and spent the next two years participating in some extraordinary research. For my thesis, I developed a mathematical model of the formation of glaciers, which was part of a Department of Energy study to determine the effects of glaciation on nuclear waste deposits. This model will be used to help select a site for a nuclear waste dump.

In 1988 I began working as a software engineer at Itek. Itek makes camera systems, airplane sensor and window systems, active control mirrors, and telescopes for government agencies. Most of the time I worked on mathematically intensive optical analysis software, but I also worked on research and development for "stealthy" airplane windows. While working on these tasks I had to

learn about optics and physics, but my background in mathematics carried me through it all. Software engineering is not just about writing computer programs. Most of the time some pretty fancy mathematics makes up the core of the computer program, and so people with degrees in mathematics are often sought after by software companies. Itek is currently manufacturing 32 hexagonal pieces of glass which will be pieced together to form the primary mirror for CalTech's Keck Telescope. Keck is a privately-funded project to build the world's largest telescope, which will eventually reside in Hawaii. The optical analysis software which I worked on is used to manufacture such optics. Test engineers collect massive amounts of data which must be analyzed to insure that the optic is meeting its specifications (so that Itek does not have a Hubble telescope incident). The mathematical algorithms used to process the data are fascinating.

While working at Itek was always challenging, I recently accepted a position as a senior software engineer at Harris Scientific Calculations in Rochester, NY. Harris Scientific Calculations produces a variety of software products, but their main product is an Electrical CAD (computer-aided design) package called Sci Cards. This next phase of my career is going to require that I learn about electrical engineering and CAD technology. This is going to present a whole new set of challenges to which I am looking forward.

Samson Cheung

BS Mathematics/Physics
London University

MS Applied Mathematics
University of Maryland

PhD Applied Mathematics
University of California, Davis

Research Scientist
MCAT Institute
NASA Ames Research Center

For centuries, mathematics has been applied to many different scientific fields, such as physics and astronomy. The field of mathematics has been extended to solve problems in computer science, highway traffic, and economics. Pursuing an education in mathematics can provide one with a wide variety of choices in business careers, while allowing one to help one's country's production industry.

After my graduation, I worked as a research scientist at NASA Ames Research Center, California. At that time my research subject was using computer codes to predict sonic boom of supersonic aircraft.

Sonic boom is a noise created by an aircraft when it flies at supersonic speed in the earth's atmosphere. As you may know, the British and French built a supersonic airplane, the Concorde, which carries 100 passengers and flies as fast as two times the speed of sound. However, this aircraft is not welcomed by airline companies, not only because it is too expensive for short flights but also because the Concorde creates an environmentally disruptive sonic boom.

Traditional methods are not always adequate in supersonic aircraft design. In order to explore viable design methodology and aircraft shapes for supersonic commercial aircraft, we need two important tools: (1) today's supercomputer and (2) mathematical optimization.

Optimization is an important branch of mathematics which is applied to solve many problems in many industries. In high school, you may have learned about the golden section search method, discovered by the ancient Greeks, which finds the minimum length of a non-linear curve. The golden section search is one of the tools used in optimization. Now-a-days optimization involves different branches of mathematics, such as linear algebra, calculus, and applied analysis. As an applied mathematician, I am using mathematical optimization to design an aircraft platform which creates a low sonic boom and provides high aerodynamic performance. However, there are more than just aerodynamic and environmental issues. In the design process, there are many constraints, such as exit doors' height, landing gear positions, and other considerations in weight and structure. We must convert this information into mathematical functions and constraints before we find the optimal solution. The most interesting thing about the optimization process is that the way one models the problem plays a significant role in the quality of the result one obtains. Proper problem modeling comes from one's experience and one's imagination.

I believe an education in mathematics can provide one with the tools to use one's imagination to help one's country.

Yves Chiricota

BS, MS, Phd, Mathematics
Université de Québec á Montréal

Postdoctoral studies, Computer Science
Laboratoire Bordelais de Recherches en Informatique

Project Manager
PAD System Technologies

Professor
Université du Québec á Chicoutimi

I have always been drawn to music in all its forms. By the end of high school, my interest focused on the electronic synthesis of sound. Up to that point I had never thought of undertaking advanced studies in mathematics, but reading several articles in a music and computer science magazine led me in this direction. The articles discussed various techniques of sound analysis and synthesis involving Bessel functions, the Fourier transform, etc. Intrigued and wanting to understand these concepts, I decided to major in mathematics.

My interest in mathematics crystallized during the subsequent years. What attracted me was its aesthetic dimension, surely as profound as that of music — though without the sounds. I gradually began to *do mathematics for its own sake*, eventually obtaining a doctorate, all the while maintaining my musical interests. I pursued postdoctoral studies in computer science and afterwards was hired by a company that makes software for the garment industry. My background as researcher and mathematician played a prominent role in overcoming the challenges which confronted me in this line of work.

One of the main problems that I tackled at this time was the development of software which allowed a computer to produce 3D images of clothes starting from their 2D patterns. Mathematics played a crucial role in accomplishing this task and in particular geometry, numerical methods and differential equations were of essential importance. This was not an isolated phenomenon. Mathematics presentented itself in all the problems I dealt with during this period. Indeed computer science turned out to be a vehicle for the realization of various mathematical objects and ultimately my knowledge of mathematics allowed for the production of particularly efficient models of real-life problems.

After five years in the industry I returned to academics. As of December 1999, I am a professor in the department of computer science and mathematics at the Université du Québec á Chicoutimi. Undoubtedly my experiences will convince my students of the fact that a solid background in mathematics is an essential ingredient for a productive and gratifying career in computer science, or other scientific disciplines for that matter.

In summary I would say that the study of mathematics, beyond the aesthetic pleasures this affords, has given me a marked advantage with regards to problem solving, analytic reasoning and the power of abstraction. These advantages have played an essential role in my professional evolution both in the computer science industry and academics.

James L. Cooley

BS Mathematics
University of Massachusetts

MA Mathematics
The Pennsylvania State University

Aerospace Mathematician
NASA Goddard Space Flight Center

What does a math major do at NASA surrounded by engineers, physicists, and astronomers? Mathematicians provide excellent background to model physical systems. The physical systems can be related to a spacecraft (an attitude system, a propulsion system, etcetera), a spacecraft support system (such as a ground or space tracking system), or a system in nature (such as the earth's gravity field or atmosphere). Mathematics provides excellent background to model data (such as noisy or biased data taken from an attitude system or a tracking system) and determine the optimal information from the data. Mathematics also provides excellent background to understand geometric relationships and deal with changing relationships over different time scales (predicting ahead, in real-time, and after the fact).

After graduating from Pennsylvania State University and adding an additional year of graduate study in mathematics at the University of Maryland, I joined the Goddard Space Flight Center in 1963. Immediately there was the challenging problem of modeling the tracking system for the Apollo program. There continued to be many challenging problems in the area of flight dynamics: the area encompassing orbits and orbit maneuvers, attitude systems and attitude maneuvers, and spacecraft tracking. Other challenges involved modeling tracking

from spacecraft (the tracking and data relay system) instead of from ground antennas, controlling spacecraft dipping low into the atmosphere, and controlling spacecraft around a mathematical point (the sun-earth calibration point). There continue to be challenges in designing future missions: tracking and controlling four spacecraft in tetrahedron formation, for example, and designing missions allowing correlation of scientific data from two or more spacecraft. Mathematics and mathematical approaches are used a great deal in modeling physical systems and designing NASA missions.

Any spacecraft mission is a team effort involving engineers, physicists, astronomers, and mathematicians. One rapidly realizes it is necessary to have a background in and learn the language of engineering, physics, and astronomy. Computer and computer science knowledge is also indispensable. Thus I always recommend some minor courses in these fields for a mathematics undergraduate or graduate student.

NASA Goddard Space Flight Center in Greenbelt, Maryland, has formed a mathematics support group to enhance mathematics and mathematics education. The group supports common areas of concern such as chaos theory (chaos theory may provide an exciting breakthrough in modeling and predicting sunspots and solar activity). This group also promotes the concept that a mathematics major who is familiar with physical sciences is as good as, if not better than, an engineer, physicist, or astronomer who is familiar with mathematics.

Mary K. Cooney

BA Mathematics
State University of New York at Geneseo

MS Secondary Education
Long Island University

Mathematics Editor
Simon & Schuster, Inc.

I imagine there are many college students today who are in the same predicament I was in when I was in college. As far back as I can remember, I was a student whose best subject was math, so, when I went to college, naturally I enrolled in math courses. I was never quite clear what I would do with a math degree, but I figured that, if I took courses I liked, eventually I would end up doing something I would enjoy.

My only role models were mathematicians who were teachers, so I assumed that teaching was what mathematicians did. I liked the idea, so upon graduation from college I immediately enrolled in graduate school to pursue a mas-ter's in education. At the same time, I worked full-time as a payroll manager. My estimation and problem-solving skills were often incorporated on the job, and computer courses I had taken in college gave me insight into the business applications of computer operations. I was able to implement an efficient method for collecting and processing payroll data on a weekly basis.

As soon as I completed my student teaching, I began teaching high school mathematics. Teaching was a wonderful experience for me. Every day was filled with new encounters. I was challenged to challenge others, to motivate

students, and to encourage them to enjoy learning. Teaching left room for plenty of creativity, and I was constantly learning new things — new things about math and new things about people.

Today, I am working for a publisher of high school textbooks, something I never even considered when I was a college student wondering what I would do with a math degree. As an editor, I am part of a team of people who develop mathematics material for use in high school classrooms. We develop textbooks, computer materials, and teachers' resource materials from concept to final product. I have had the opportunity to travel to several cities throughout the country to explore trends in math education and to meet and work with several authors and teachers. I have been introduced to the technological industry of desktop publishing, and, in the process, I have learned several computer software programs. Although I am no longer teaching, I still tutor high school and college students on a part-time basis. My math background has even afforded me an opportunity to write test items for a well-known college testing service.

Recently I enrolled in several chemistry courses, and I am exploring the opportunities afforded by a dual mathematics/chemistry degree. Although I am not certain of exactly what opportunities and challenges await me, I am eager to explore them one day at a time.

Jeff Cooper

BA, Mathematics/Computer Science

Gettysburg College

MS, Telecommunications
Johns Hopkins University

Corporate Technical Developer
P.H. Glatfelter Company

Since 1992, I have worked on projects to streamline the business practices for a leading paper manufacturing company. My current job is to maintain knowledge of new and emerging technologies in the computer field that can improve the bottom line of the company. I maintain areas of local and wide area networks that support manufacturing facilities throughout the United States. I installed Internet connectivity and email for the entire corporation, and have assumed the Webmaster position. This project gave me the opportunity to work with everything needed to support a World Wide Web site.

I started with the company in 1992 as a computer programmer. I began my career writing business systems from written specifications for an IBM mainframe. I have used many computer-programming languages such as COBOL, FoxPro, C++, PowerBuilder, Pascal, and JAVA to write computer programs. In this position I used my mathematics degree to develop algorithms that would maximize CPU utilization and write efficient computer programs.

After six months, I was promoted to a Systems Analyst. In this position I designed computer software for business systems such as accounts payable, time management, steam collection from turbine generators, sales systems, inventory management and many others. I used my mathematics knowledge every day for compiling

statistics and calculus and numerical analysis for modeling complex software systems. I was on the forefront in developing client-server applications, and implementing new Information Systems architectures with my company.

After two years as a Systems Analyst I was promoted to my current position. As a Corporate Technical Developer, I must be aware of new and emerging technologies. I often travel to trade shows such as COMDEX, Comnet, and other large shows throughout the country to evaluate the new technologies. My corporate responsibilities require me to support computer systems at three manufacturing facilities across the United States. I often travel on our private jet to these sites for an extended or just a day visit to implement these new technologies.

I am also the Webmaster for the Glatfelter Company. This has allowed me to build the Internet infrastructure from the ground up. I installed the hardware and software necessary to support a WWW server, Internet news server, email server, intranet server, and other Internet hardware and software. Another part of this job is to develop the actual web site and all the web pages. This has been a rewarding and challenging task; all made possible by my mathematics degree. My career at Glatfelter began with a solid foundation in mathematics. My job allows me to actively use some of the leading tools in technology and putting them to work in the business world.. Since then I have been working to obtain an advanced degree at Johns Hopkins University. Having a mathematics degree opens the door for almost any career you can imagine.

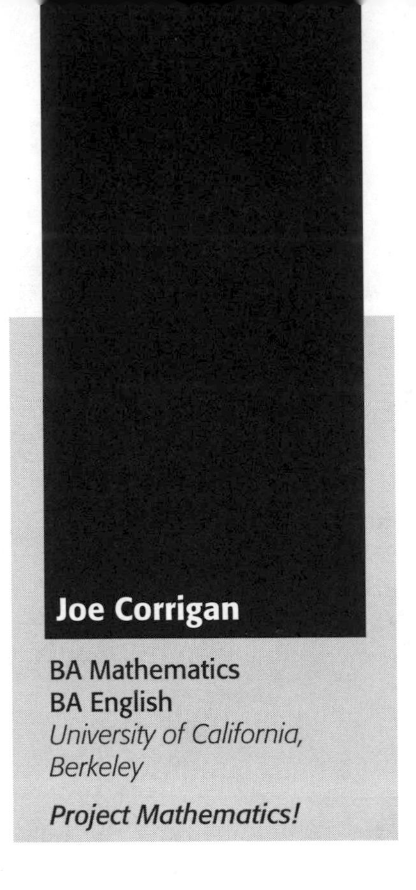

Joe Corrigan

BA Mathematics
BA English
University of California, Berkeley

Project Mathematics!

As I reflect upon the many interesting projects I have been involved with over my career, I am struck by the fact that in almost every case my background in both mathematics and English has been an invaluable asset. As I have discovered, my stubborn refusal to give up one interest in favor of another has allowed me to undertake some intriguing tasks.

Fresh out of school, I went to work as a Public Affairs Officer at the Naval Medical Center in Bethesda, Maryland. The ability to write and edit coherently was certainly an asset, but, without my mathematical skills, I would not have been effective in covering the numerous areas of scientific research that was my "beat." My next job was with a film production company in Hollywood, where I became involved in the first use of videotape in television commercials. The introduction of video recording technology opened up for me an entirely new area of technical exploration. But, without my writing skills, I would have missed it completely.

In 1979 I joined a project to broadcast a closed captioning system to enable hearing-impaired viewers to enjoy television along with the rest of the population. Aside from a thorough knowledge of broadcast video systems, I relied heavily on my math background in building the technical data creation, encryp-

tion, transmission, and decoding systems. And I relied on my English skills to establish computer-based language coding, timing, and translation systems. Next, I turned my attention back to the film industry. Through the use of subtitles and soundtrack dubbing technology, the entire world has long enjoyed Hollywood motion pictures. But, until the introduction of video and computer technology, the production of subtitles and foreign language soundtracks was a costly, time-consuming process. With my background, I welcomed the challenge of creating new computer/video language translation and generation systems.

Now I am working with *Project Mathematics!*, an award-winning series of instructional modules that use outstanding computer animation, age-appropriate video and film footage, and carefully crafted narration to assist instructors in teaching basic concepts of high school mathematics. The videos and workbooks present mathematics as exciting, challenging, and socially acceptable. My work with *Project Mathematics!*, which is produced at Caltech, has allowed me to once again fully utilize my mathematics and English skills without having to choose which I like more than the other. For me, the choice is obvious: I want the best of both worlds.

H. Michael Covert

BS Mathematics
The Ohio State University

Senior Computer Systems Specialist
Chemical Abstracts Service

My interest in mathematics started early in life. I showed an aptitude for the subject in elementary school and by junior high school had started accelerated course work in algebra and plane geometry. The high schools that I attended were fortunate to have computers as early as 1970. My real commitment to mathematics was cultivated by one exceptional high school teacher. He set up a small study group that supplemented normal study activity with material put together by each study group member. The experience was one of the most valuable in my life since it taught me to work with a team. We covered an extremely wide variety of topics ranging from elementary calculus to tensor analysis.

My college experience was atypical. At Ohio State University I completed nearly all of my mathematical course work by the end of my sophomore year and saved my basic educational requirements until my junior and senior years. In retrospect, I believe that this was a good way to proceed since my mathematics course work organized my thought, making liberal arts course work "easier." It also provided a good foundation for delving into physics and astronomy since my mathematical training facilitated an understanding of those subjects.

In my sophomore year I obtained a position with the Instruction and Research Computer Center (IRCC) at OSU. I began by servicing printers, keypunches, terminals, etcetera. Then I moved to consulting and finally into systems programming. I

used my mathematics background in nearly every aspect of my job. I wrote the OSU Neutral Plotting Code system, a device-independent graphics system that utilized linear algebra, differential calculus, curve fitting, etcetera. I did extensive work in behavioral sciences that enhanced my mathematically based understanding of statistics. Eventually my systems programming led to some limited computer performance analysis using simulation and analytical models.

After graduation I accepted a position at Chemical Abstracts Service (CAS). CAS provides comprehensive chemical information to the world. This information is provided in the Chemical Abstracts journals and within the Scientific and Technical Network (STN) online service. Currently I am the project leader within the operating systems group responsible for maintaining multiple mainframe-based computer systems. My mathematical training was essential in providing the skill necessary to attain this position. Typically my duties involve planning, technological evaluation, providing performance and capacity planning services, and emergency support for these systems. CAS has offered me an education that has all the benefits of academia combined with "real life" commercial experiences. CAS maintains a commitment to current technology; thus the job never becomes boring or repetitive.

I can summarize my experience by stating that by starting with mathematics I made the best possible choice. Other disciplines, notably computer science, were possibilities, but the basis for nearly all science is mathematics, and as such it provides a platform for an extremely wide variety of career options.

Sarah Cullen

BA Mathematics
Potsdam College of SUNY

Senior Professional Relations Representative
Bay State Health Care

If you are considering a major in mathematics in college, have you ever wondered what kind of job you will be prepared for when you complete your bachelor's degree? I did, and I was still wondering when I walked across the stage to receive my BA degree from Potsdam College in upstate New York in May 1986.

When I missed most of the companies who were visiting the Potsdam campus to recruit in the fall of my senior year because I was studying in England that semester, I knew I had to take my own course of action to find a job. I have always loved the ocean, so I moved to the Boston area and began answering ads in the Boston Globe for positions that sounded interesting.

After being a finalist for a job I really wanted and didn't get, and crying my heart out for a few minutes, I sent my resume and a cover letter to Bay State Health Care for a financial analyst position. In competition with some applicants with MBAs, I was interviewed by several people, shown around the company, and eventually offered the job.

At first I spent most of my time working on a personal computer, setting up spreadsheets and doing some analysis. I assisted with premium-rate development for our marketing groups involved with the health maintenance organization. I also interacted closely with the management information systems area to

design and tailor reports for use in our department. Fortunately, I had taken computer courses in college, and had worked part-time in a hospital in the Finance Department with computers. I was also fortunate in having a wonderful supervisor who helped me get started on the right foot.

After I had been with Bay State for a year, I had an opportunity to work part-time with Professional Relations, visiting physicians and their office staffs. I enjoyed the work, and continued to be a Professional Relations representative for eight years. I was responsible for a territory of 800 physician offices. I conduct on-site educational visits, serve as a liaison between the physicians and Plan management, recruited physicians, and trained offices on new communication technology.

I am now a Provider Relations Manager with the Northeast PHO at Beverly Hospital. I have two children, and I am able to work three days per week. I continue to act as a liaison with physician offices, conduct educational programs, draft a monthly newsletter and support a clinical committee.

I am not using calculus or any other specific math course I had taken, but I am using the precision of language and the analytical and problem-solving skills resulting from completing my math major in college. Perhaps there is a career opportunity waiting for you that you have not yet even considered.

Allison DeLong

BA Mathematics
Grinnell College
MS Mathematical Modeling
Humboldt State University
Marine Research Associate
University of Rhode Island

Fish provide an important source of protein in the U.S. and the world. We must manage the oceans well so that we can maximize the amount of fish that can be sustainably harvested. History has proven how challenging it is to do this. Several populations of the most desirable fish species have been reduced to very low levels from commercial exploitation; for example some populations (and subsequently catches) of haddock and flounder off the northeast US were severely reduced in the early 90s. Several Pacific salmon populations are considered threatened and endangered. Scientists and resource managers are trying to determine how to rebuild these populations, and even whether it is possible to do so. The people who look into these questions have a solid understanding of such topics as ecology, biology, statistics, dynamical systems, and linear algebra—just to name a few. It is very common for individuals with backgrounds in both applied mathematics/statistics and biology/ecology to pursue this field.

Since June 1997, I have been working as a Marine Research Associate at the University of Rhode Island's graduate school of oceanography. I work on projects related to fisheries stock assessment and population dynamics of exploited fish populations. I am funded through grants from such institutions as the National Marine Fisheries Service and state agencies. These agencies are charged with sustainably managing the aquatic resources and habitat in the areas under their jurisdiction and our findings aid them in their mission.

One of the projects I recently completed was to improve the model used by the Alaska Department of Fish and Game to estimate several king crab population

abundances. The Department permits the retention of crabs that are greater than the size at sexual maturity, so the crabs will have an opportunity to reproduce at least once. Managers set a commercial quota each year to be a proportion of these legal-sized crabs, if the population biomass is above a threshold value. If it is below the threshold, the fishery is closed for the year. Furthermore, they do not allow the quota to exceed a set proportion of the mature population. This management policy requires estimates of both the mature and legal populations. The previous model was not able to estimate the mature catch, so we extended the model to do this. We then tested and compared the new model to the earlier one with Monte Carlo simulations and other statistical tests. This new model is now used to help set the quotas for several king crab populations.

Another ongoing project is to develop a dynamic model that describes population growth and decline as a function of both commercial exploitation and predation, with the hope of better understanding this relationship. We are examining the mathematics of the resulting model to determine if there are multiple stable equilibria and whether this may influence the ability of the system to return to its pre-fished state.

I chose to pursue this field in graduate school when I learned how interesting quantitative ecology can be and how vital this information is to regulators as they attempt to manage resources. The ocean presents another level of complexity, as it is difficult to observe the resource being managed. I love to apply mathematics in this way because I collaborate with intelligent individuals to answer interesting questions and because the challenges are endless.

Janet P. DenBleyker

BS Mathematics
Bucknell University

Assistant Actuary
Buck Consultants

What is the probability that the roulette wheel will stop on 33 black? What is your expected payoff if you bet $100? How long are social security benefits expected to last? These are the types of problems actuaries can solve. The job of an actuary is to assign probabilities to expected events and determine the expected cost associated with the outcome. Actuaries are involved in determining the expected wealth of your pension plan, developing a medical benefit package for your company, calculating the amount you save on auto insurance by installing an anti-theft device, and advising your mayor on the costs of providing care to the elderly in your city. There are actuaries in all areas of the business world: insurance companies, consulting companies, investing companies and government organizations. I selected a career at a consulting company in the health and welfare department.

At my job, we are involved in all aspects of an employer's benefit package, from start to finish. We help the company decide what types of benefits they want to provide and how much each benefit will cost. We help the company select an insurance company to insure the benefits. We price their renewal rates and negotiate with the insurance company when it is time to renew the policy. When a change to the plan is needed, we price the cost and savings associated with adding, changing and eliminating certain benefits.

We also provide consulting advice to insurance companies and those providing the insured benefits. We help determine the level of funding needed to pay out the expected claims. We price the rates to be charged for new insurance products, keeping in mind the company needs to remain competitive in the marketplace yet charge enough to cover all the expected expenses. Additionally, we help providers manage the high costs of healthcare and provide benefits to those that need healthcare but can't afford it.

In addition to the daily functions of my job, as an actuary I continue my education by taking exams administered by the Society of Actuaries. The tests are challenging but should not scare a candidate away. Most employers provide their actuarial students with the time and resources necessary to help a potential actuary successfully make their way through the testing process.

My advice to someone interested in the actuarial profession: talk to many actuaries in different areas of the industry to get a feel for what is right for you, pay attention to the constant changes in the healthcare and insurance industry and start taking the tests. A career as an actuary can be very interesting and rewarding.

Mark Derwin

BS Mathematics
MS Operations Research
SUNY at Stony Brook

All through my high school years I learned all that I could about computers. By the time I enrolled in college, I knew three computer languages and three computer systems. I figured that computer science was the only way to go. I was wrong. By the end of my sophomore year, I was bored and looking for new challenges. That was when I discovered applied mathematics and statistics.

After two years of calculus and programming, I finally had a chance to apply some of the theory. New ideas and challenges were presented to me. I particularly enjoyed puzzles and problem–solving. Courses such as graph theory, game theory, and combinatorics provided me with new insight for solving difficult problems.

At graduation I decided that the real world was not ready for me. I applied and was accepted to the Stony Brook Masters program in operations research. These courses taught me how to budget and organize my resources, including time. Also, I learned to apply my skills to real-life situations. No longer was everything theoretical. One of the most rewarding parts of my MS study was a practicum study of the Long Island Railroad on railroad car utilization.

During my second semester of graduate school, I started working for a small software firm as a technical writer. While learning and documenting their sys-

tems, I discovered many bugs, and I was promoted to Software Quality Assurance. They actually paid me to break systems and find bugs!

Now, three years later, I'm at a much larger firm, and I'm still breaking systems. Many of my problem-solving skills are used to break through security. I'm sent all over the country to give courses and classes on the systems I use and the methods I employ. I've come a long way from being afraid to speak in front of a small group of students and faculty.

One suggestion that I have for all students, both graduate and undergraduate, is to find a professor to whom you can talk. An advisor can help steer you in the proper direction by helping you identify your strengths and weaknesses.

Looking back on college, I am appreciative that my applied mathematics courses taught me general strategies for problem-solving as well as specialized mathematical techniques. Another important aspect was that I learned how to work in small groups and hash out difficult problems. For now, I plan to continue breaking systems, and then I may go into building them. There is also a good chance that I'll continue teaching and try to pass along some of the information that I've learned.

Susan J. Devlin

BS Mathematics
William Smith College
MS Statistics
Rutgers University
Director - Measurements Research
Bellcore

My professional career, first at AT&T Bell Laboratories and then at Bellcore after the breakup of the "Bell System" in 1984, has had two distinctly different phases. Both heavily utilize my graduate training in statistics.

When I completed my degree in mathematics in 1968, I knew I wanted graduate training in a mathematical science, but I felt ill-equipped to choose a specialty. Thus I went to work for AT&T Bell Laboratories as a senior technical assistant in the statistics and data analysis research lab, which offered exposure to applied mathematics, statistics, and computer science. Bell Laboratories offered the opportunity to work with researchers in methodology development and participate in diverse applications. Going to school part-time, I completed an MS in Statistics and all course work and exams for a PhD.

The 17 $^1/_2$ years I spent in statistics research were exciting and filled with opportunities for professional growth. I did basic research with diverse teams of statisticians and occasionally with other scientists, published over 20 papers, and prepared numerous internal technical reports, covering such areas as demand analysis, regression diagnostics, robust estimation, seasonal adjustment of time series, and multivariate smoothing. I also consulted on interesting studies, such as predicting telephone demand and price elasticities, optimizing the selection of cities for expanding video conferencing, and modeling the fading of radio signals based on weather and geographic characteristics. I embedded some of our developments into corporate statistical training and lectured at our corporate training center.

With strong support from my employers, I have been active in the American Statistical Association (ASA), the Caucus for Women in Statistics, and the Mathematical Association of America, sitting on committees, holding offices, and making technical presentations. In addition, my employers' time and resources allow me to teach mathematics one hour per day in an inner city school, to help manage a national lectureship program, and to conduct analyses to assess the impact of race and sex on statistical salaries and participation in ASA activities.

In 1987 I made a major change in my career direction: I left research to take over the development of a multi-disciplinary group to study customers' satisfaction with the service provided by Bellcore and its client companies. We develop models and methods for assessing telecommunication products from the perspectives of consumers and businesses. We also evaluate Bellcore products and services: large software systems, network design, technical consulting, training, and corporate support functions (such as secretarial support, building security, and cafeteria services). I supervise a staff of 15, travel extensively to plan work or present results to our clients, and consult directly with our officers. Our work is utilized to direct quality improvement efforts and as input to management compensation plans.

Although I miss performing my own statistical analysis and have less time to publish, the tradeoff has been to develop and plan research strategies, expand my knowledge through the diverse skills of the members of my group, and gain expertise in budget administration, personnel development, executive-level presentation of technical ideas, and product management/sales. What is most rewarding is to see our work immediately utilized by our clients.

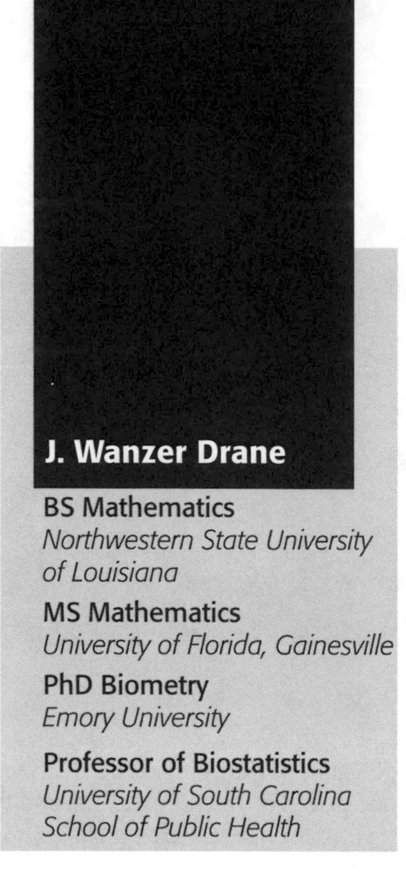

J. Wanzer Drane

BS Mathematics
Northwestern State University of Louisiana

MS Mathematics
University of Florida, Gainesville

PhD Biometry
Emory University

Professor of Biostatistics
University of South Carolina School of Public Health

Can you imagine a researcher and teacher in biostatistics being involved in 1) Creating a NIGHT WATCHMAN, 2) Revising CLUSTER, software used for investigations of disease clusters, 3) Modeling the stochastic nature of a progressive and irreversible disease, and 4) Applying the methods and mindset of biostatistics and biometry to the repair and maintenance of army helicopters? These are four areas of research that occupy my time and attention at present.

THE NIGHT WATCHMAN software is used to analyze data on reportable diseases. One can envision its structure by creating a matrix of disease frequencies and their severities. At each intersection a statistical program is called to analyze the appropriate data for its spatial and temporal characteristics. Too many cases close in either space or time would signal an alarm. At this writing NIGHT WATCHMAN can best be described as developmental, but with realistic expectations of implementation in the near future. It is being developed in collaboration with the Tyneside health district, covering the area north of the River Tyne to the border of Scotland.

CLUSTER has been in the public domain since 1993. A collaborator in Verona and two of us here are adding enhancements to capture a wider range of situations, some of which might also be included in THE NIGHT WATCHMAN. Much modeling remains to be done to link environmental hazards with chronic diseases and syndromes. CLUSTER does not provide the link between environ-

ment and disease, but it is used to analyze case and grouped incidence data over arbitrary intervals and space. From its use one can develop hypotheses connecting environmental factors with diseases for later investigations.

Progressive diseases can be modeled using three disease states and random waiting times between successive states: Disease Free, A-Symptomatic, and Symptomatic. Each of these states has its own waiting time distribution for progression to the next and more severe state. It is possible to model the three epochs using trivariate distributions on the nonnegative reals. The ever-popular trivariate normal will *not* satisfy the requirements of the problem. Simulation and bootstrapping ought to advance our knowledge in this domain.

Complex machinery such as helicopters can be modeled as Disease Free, A-Symptomatic and Symptomatic or simply ready for an overhaul. For the human, laboratory test are used to detect presence of out of normal parameters such as blood chemistries and disease specific tests which are especially recommended for older men and women. With a helicopter or another complex machine the same approach can be taken to measure progression from health to illness or from readiness to needing an overhaul.

I encourage young mathematicians who are leaning toward applications to take courses to increase their knowledge base. Is my work ever finished? I hope not! Research uncovers more problems than it solves. Life continues to be more interesting with each day, and there is never a boring moment for this biostatistician.

Arnitra Duckett

BS Mathematics
Spelman College

Educational Markets Manager
Texas Instruments, Incorporated

Throughout my educational career, I found mathematics to be a challenging and stimulating subject. The thrill of problem solving and linking those solutions to other mathematical phenomena peaked my interest. After completing high school, I enrolled at Spelman College and decided to study mathematics. Throughout those four years, I not only became proficient in the language of mathematics, but also received a well-rounded liberal arts education.

Upon receiving my Bachelor's of Science degree in May of 1995, I began my career with Texas Instruments, Inc. as an Educational Markets Manager (EMM). As an EMM, my job is to create awareness and demand for classroom technology. In doing so, I am able to address the needs and concerns of mathematics and science educators throughout the Mid-Atlantic United States. I provide educational product and support program information to these educators as well as inform Texas Instruments of the educator concerns which I may encounter. I also assist mathematics and science teachers with the incorporation of calculators into the mathematics and science curricula as a learning tool.

During my first year with Texas Instruments (TI), I have been able to interact with mathematics and science educators on many levels. Through teacher professional development sessions and regional conferences, sponsored by TI, I am

able to aid educators with the use of a calculator as a classroom tool according to district, state, and national mathematics/science education guidelines. Because of my mathematical background, I am able to discuss and understand pedagogical issues with educators so that technology can be implemented in a useful manner.

Not only do I co-organize technology-based conferences, I also attend local, regional, and national mathematics and science conferences. My attendance, along with the attendance of my TI colleagues at conferences, such as Mathematical Association of America (MAA) and National Council of Teachers of Mathematics (NCTM), give several of the educators attending a "warm, fuzzy" feeling inside and brings meaning to the phrase "TI-CARES."

I have always been aware of the connection between mathematics, science, and many other subjects. However, the connection between mathematics and business became more apparent as I began working for a major corporation. My experience with Texas Instruments has allowed me to understand exactly how powerful mathematics is in today's society and brings the phrase "Mathematics is Everywhere" to life. Because of my involvement in both the mathematics and business communities, I plan to pursue a Master's in business administration in order to strengthen my bond between mathematics and business.

Paula Duckett

BA Sociology
George Washington University
MA Education
Catholic University
Elementary School Mathematics Specialist

I am a member of an important yet often overlooked group of professionals — that of elementary school teacher. It is the elementary school teacher who gives a child his/her first formal learning experiences, and these experiences have a profound impact upon the child's success or failure in school.

Because of the strong emphasis now placed on the improvement of mathematical sciences education, the role of teaching on the elementary level is being altered to allow teachers to develop skills in one or two subject areas. This trend is currently evident in elementary mathematics education. A new title is slowly evolving — the elementary math specialist. A math specialist is not a curriculum writer, or a mathematics supervisor, or even a resource teacher, but an elementary teacher who teaches mathematics and only mathematics in the local school. The role of the math specialist is to instruct, diagnose, remediate, and evaluate the progress of the heterogeneously grouped students that comprise each class. This responsibility relieves the regular classroom teacher of the task of teaching mathematics while simultaneously allowing a specially trained teacher to focus on one subject — mathematics.

I enjoy the challenge of planning new and interesting ways of helping my students discover the magic, the mysteries, and the power in the patterns of mathematics. It is exciting to present new mathematical concepts and watch as students

experiment and test various mathematical hypotheses in order to reach a conclusion about ratios, percents, or whole number operations.

What fun it is to determine the percentage of a banana that is edible; or estimate how many meters you can throw a "javelin" made of a plastic straw; or find the average number of peanuts in a candy bar; or use colored rods to assist in the addition or subtraction of unlike fractions; or model square and triangular numbers using ceramic tiles.

I recommend a career as a math specialist to anyone who enjoys teaching mathematics and welcomes the challenge of helping youngsters develop mathematically. My math lessons may not be structured around differential equations or calculus, but the concepts I teach are just as important to the elementary math student. I try to instill in my students my enchantment with mathematics and an appreciation for the many worlds that they can own by mastering mathematics. The students who leave my classroom are empowered to be successful in mathematics.

Thomas E. Dunham

BA Chemistry
Ohio Wesleyan University
PhD Metallurgy
The Ohio State University
Vice President and General Manager
Service - GE Medical Systems
General Electric Company

I was trained to become a research metallurgical engineer but, after a few years with GE, I left basic metals research. I moved from lighting products to appliances to medical systems and worked in manufacturing, engineering, and service.

My early goal was to be a materials scientist. I studied metallurgy, physics, ceramics, and mathematics, and received a PhD in 1968. My PhD thesis was experimental, and it required me to use my training in mathematics and computer science to analyze data.

I don't use much of my specific scientific training today, but I continue to rely heavily on my understanding of mathematics.

I began working for GE in the lighting business as a materials research scientist, but, after a few years, I became impatient with the slow pace of the research work and changed to manufacturing as a plant manager. Although the products we were making were new to me, I found that I could use my mathematics background to improve quality through statistical process-control techniques.

A few years later, I moved to GE's appliance division, where I led the refrigerator, range, and laundry engineering efforts of over 500 people. Once again my mathematics background helped me to understand the sophisticated modeling techniques used to test new products in the computer before they were built and released to customers. Using these techniques, we frequently predicted the quality and performance of products without having to build prototypes.

Because engineers are creative people who have lots of ideas, there were always more ideas than funds to develop them: the real question often became which programs to fund. Mathematical modeling of costs and benefits over the projected life of a new product frequently provided the information to help us answer this difficult question.

In 1988, I moved from the appliance division to GE Medical Systems, where we manufacture x-ray, CT, nuclear, ultrasound, and magnetic resonance imaging equipment. At medical systems I had responsibility for manufacturing and information systems before moving to my current position as Vice President of Service. Once again, my mathematical background helps me to analyze opportunities, forecast future resource needs, and predict results.

Students often ask me what to take in school ... my answer? ... follow your natural interests, but take lots of math!

Yawa Duse-Anthony

BA Mathematics
Colby College
MS Industrial Engineering
University of Massachusetts
Senior Associate
KKO and Associates

Most people have never heard of transportation planning, or if they have, they equate it with what travel agents do for cruise lines and airlines. However, what we do at KKO has very little in common with travel agents. We answer the questions "if we build it will they come", and "how can we get more people to use transit" for passenger railroads and transit systems throughout the United States.

KKO and Associates is a small consulting firm specializing in transportation planning and management consulting for transit authorities and private companies. Examples of projects I have worked on include the operational analysis and costing study for the Lowell-Nashua commuter rail extension for New Hampshire DOT, the development and adjustment of commuter rail schedules on the MBTA Old Colony lines to anticipate new service on the Greenbush Line, the development of a thirty year rolling stock and planning vision for the former New Haven Railroad in Connecticut which involved building a model to track equipment requirements and ridership forecasts, and the audit of mechanical failure and accident reporting for bus operations at the MBTA. I am currently working on evaluating the model outputs for ridership forecasts for the Hartford-Manchester/Vernon Bus Rapid Transit Feasibility Study for CRCOG and the Connecticut DOT and developing improvement packages for the Worcester Service

Expansion Study These projects illustrate the variety and breadth of the work that we do at KKO.

The common thread in all the projects I listed above is the analytical background required to understand and develop solutions to the problems faced by transit agencies and other companies concerned with moving people by rail or by bus efficiently and cost-effectively. The math degree I received at Colby College plays a huge role in my ability to comprehend complex tasks and build economic and forecasting models, as well as provide analysis of data that comes to me from a variety of sources. A math education teaches you not just math, but the ability to learn and to apply what you have learned to other tasks that at first glance may not seem to be related or to have anything in common.

My math education also helped me greatly in graduate school. When I entered the masters program I already had a firm grip on the mathematics behind engineering, and I had the fundamentals of learning and working in teams down cold from the environment fostered among math students of cooperative effort and group study. I think that because math fosters team building, math students are more prepared for the work place and for graduate school than most other majors.

M. Scott Elliott

BA Mathematics
University of Virginia

MS Operations Research & Management Science
George Mason University

Operations Research Analyst
FedEx Corporation

My job entails using mathematical tools to create worldwide (short-term, medium-term, and long-term) computerized forecasts that measure fluctuations in aircraft ramp volumes.

My interest in mathematics and computers began as a young child. I enjoyed the challenge of solving puzzles — everything from jigsaw puzzles to brain teasers to word problems. Throughout my undergraduate studies, I specialized in mathematics. However, it was not until my last semester that I was exposed to operations research. During this course, I was introduced to using mathematical models and computers to solve real-life problems.

One of the problems that I was asked to solve was how to minimize the waiting times for individuals who want to ride Space Mountain at Disney World. Utilizing queueing theory, I was able to evaluate how dividing a line of people into one, two, three, or four smaller groups would affect waiting times throughout different times of the day based on crowd size. A second problem that we examined was determining how a farmer might maximize potential profit. Using linear programming, we were able to determine which crop (corn, wheat, or beans) would offer the greatest revenue after accounting for fertilization and harvesting costs.

I was so inspired by that single course in operations research that I went on to obtain my master's degree in OR. I was already employed by FedEx as a senior service agent when I began my graduate work. Halfway through my masters degree program, I was relocated by FedEx to Memphis, Tennessee, in order to work on their worldwide operations research department as an operations research analyst. The company allowed me to commute two days per week between Memphis and Washington, DC, in order to complete my studies. During the two years when I was completing my course work, significant strides were made in the computer field enabling larger mathematical models to be solved in a fraction of the processing time. The creation of network models and forecasting models significantly enhanced the problem-solving capabilities of vast areas of the computer industry.

Looking back over the past seven years, I could never have imagined that my interest in mathematics would take me so far, so fast. The strength of my mathematical background has been very beneficial in helping me achieve my career goals. There is great satisfaction in knowing that every day I work, I am able to solve problems that are interesting, exciting, and challenging.

As a Civil Engineer a large part of my job consists of calculating and analyz ing various situations. Some of these include (1) the cost to build a resi dential subdivision or housing project, (2) the amount of dirt required to be excavated, and (3) the amount of storm water runoff (rain) a project can hold before flooding occurs.

The analyses I regularly perform range from sizing pipes for water distribution and sewer collection to determining the required horsepower to pump storm water out of farm fields before the crops die.

The best part of having a mathematics background is that I am able to make a difference in other people's lives on a daily basis. Without my math foundation, I would not be competent to design roads, drainage systems, flood relief projects or water and sewer systems.

My first calculus teacher emphasized during each class, and I quote, "you need to know this stuff'. I did not realize the huge part calculus plays in everyday life. Calculus may not help you with wrapping your Christmas presents. But, if you wanted to know what shape to make the package to minimize surface area and the raw materials to make the package (and thus save money if you were manufacturing the package) then calculus is the tool you need.

I want to be honest with you, math is a tough subject. Mathematics can be as boring or interesting as you want to make it. I have recently financed a new car. Using my math skills, I performed a cost analysis to compare several lending institutions to determine which was offering the most cost effective deal. I used a computer spreadsheet program (EXCEL) to compare the down payments, interest rates and length of the loan. What I determined was that I could save a lot of money by getting a 3-year loan instead of a 4-year loan. When I did some research I discovered banks drop their interest rates when a customer wants to pay off the loan before the maturity date. My math skills helped me make an informed decision without relying on a "slick" salesman to interpret the numbers.

To wrap this up, my mathematics tools help me make life better for my community, they help me make wise choices as a consumer and they give me a valuable skill that ALL employers want. The last point I would like to drive home is that everyone capable of reading this article can do what I am talking about! I cannot tell you how many times I thought, "oh, I am too dumb to understand this", and thought about quitting my career in math. What I have learned in my 30 years on this earth is that anything worth having takes a mountains' worth of effort. Follow me and we will move mountains!!

Helaman Ferguson

AB Liberal Arts
Hamilton College

PhD Mathematics
University of Washington

Sculptor

Mathematics and art: Doesn't it sound as though these are two things that don't go together? When I talk to art people, they usually apologize, "Math was my worst subject," and, when I talk to math people, they tend to complain that they "can't draw." The joining of art and mathematics is possible for everyone who touches and is touched by my sculpture.

Mathematics and its uses are complex. Among other things, mathematics is a language that has three interesting features.

Feature A: you can choose a level of abstraction (eliminate inessentials).

Feature B: you can economize (condense information).

Feature C: you can predict a lot of what will happen (control the future).

These three features were crucial in my discovery that mathematics is a great design language for doing sculpture.

Let's take a simple example and show how these three features of mathematical language come up naturally. The example is a sculpture that I carved in stone — a pair of Klein bottles that link and unlink (see p. 56.) When linked, these two rotate around and through each other. To link and unlink they translate through and past each other. It is remarkable to me that two pieces of stone could have such a relationship. Relationships are important for everybody. Human

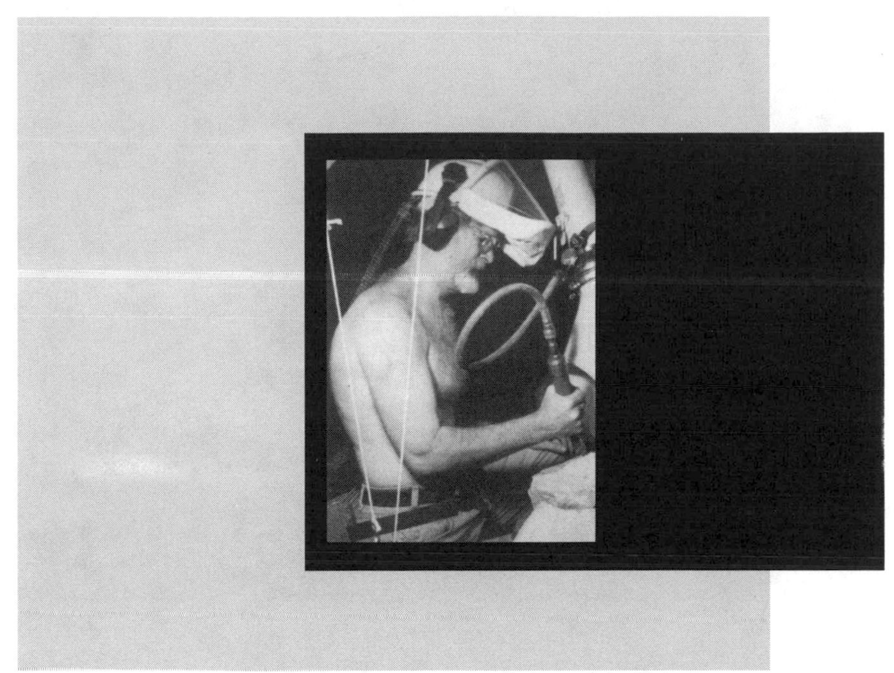

Photograph by Claire Ferguson

relationships have been an expressive subject for visual artists for centuries. The 32-year relationship between my wife and me was the relationship I had in mind while I was discovering this sculpture. Of course, every such relationship is very complicated and involves many elements. How do I say something vital and expressive about this with two rocks?

Feature A comes into play. I abstract the essential idea of orbiting each other's psyches, coming together, parting, returning. I abstract these ideas to geometric rotation and translation in three dimensions. Now I have a chance of saying something with two rocks. Many elements of a relationship have been eliminated, but with this mathematical language I have consciously and deliberately made a choice of a level of abstraction. This is a basic step in applying mathematics. Feature A helps get the chaos of human relationships organized in some way in my head and frees me to use the sub-language of topology.

Now that I have this coherent kernel of thought with a structured language, I can use some topological ideas to prove each of these things in a Klein bottle — this thing being like a torus with two plane intersections at right angles to form a surface of two Moebius bands joined together. Now I employ Feature B: develop equations for lines, planes, circles, and tori. Such relatively simple equations are very economical and summarize a zillion cubed locations in space.

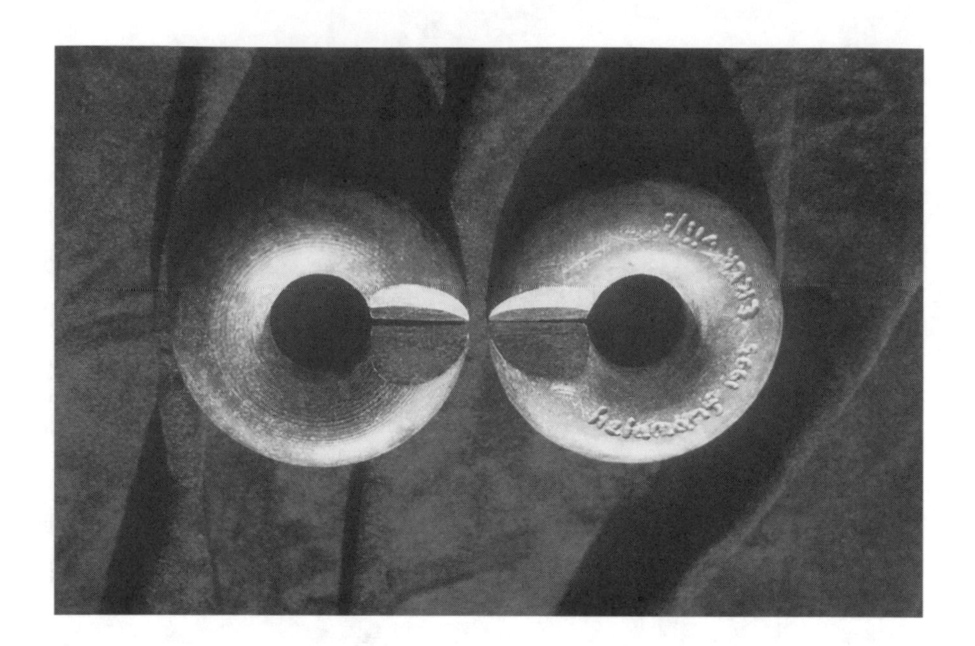

Two Carved Stone Klein Bottles that pass through each other, link, and rotate around one another.

Photographs by Jonathan Ferguson; permission for their use granted by Helaman Ferguson.

From all these equations I can generate lots of computer graphics images. Computer graphics is all fake but is a half step into physical space. I am starting to use Feature C, probing the future: these pictures give me some idea of how things will turn out physically. Stone has volume and mass, so I calculate the volume of the enveloping torus using integral calculus, a very specific prediction. But I want the two Klein bottle surfaces to be as touching and tangent as possible, so as to get them as close as possible when linked. So I solve a little minimum problem, back to differential calculus, and, again, back to Feature C. I am using my abstract design language to predict the possibilities, sizes, and behavior of this sculpture. So, when I finally cut the rock everything fits together for long life and happiness!

Rol Fessenden

BA Mathematics
SUNY, Binghamton

MA Geology
SUNY, Binghamton

Director of Inventory Control
L.L. Bean

Have you or anyone in your family ever placed an order with L.L. Bean? You might be surprised at how difficult it is to ensure that we have your specific item, color, and size when you order it.

You may remember in 1988, on Sunday before the Republican primary in New Hampshire, that the pollsters predicted Bob Dole would win. Two days later, George Bush won instead, and the rest is history.

At L.L. Bean we have thousands of items, and tens of thousands of colors and sizes. Nine months in advance, we have to predict for each one, how many our customers will want. If professional pollsters can be wrong about the New Hampshire primary two days before it occurs, imagine how wrong we can be nine months in advance on 50,000 predictions.

I graduated in 1969 from the State University of New York at Binghamton with a BA in Mathematics, and in 1974 I received an MA in Geology. Before coming to Bean, I was a geologist, a scientific systems designer, and a small business manager. In all these occupations mathematics played a key role. I developed models for surface runoff, models of heat transfer in complex, high temperature

environments, and statistical profiles of accounts receivable. I have used sophisticated statistical procedures to gain insight into glacial landforms.

My current interests are in customer testing and in mathematical modeling of the merchandising process. In mathematical terms, there are clear analogies between inventory management and previous areas of my expertise such as heat transfer. The insights I have as a result of previous work have proven invaluable at Bean. Mathematics has provided me a common framework for understanding concepts in many fields.

At Bean we attack the forecasting problem at every stage of the process. We analyze historical sales data, model forecast uncertainty to design contingency plans, and analyze catalogue displays to understand how some can be more successful or attractive than others.

We also survey customers to find out their preferences among the products displayed. On some products, the customer feedback is an accurate indicator of sales, but on others it is not. We have to understand when the questionnaire is accurate, and when it is not. Finally, we analyze the first orders received from newly mailed catalogues to refine our predictions.

Next time you order Blucher Moccasins in size 6 narrow, think of me.

Frederick L. Frostic

BS Mathematics
United States Air Force Academy
MS Engineering
University of Michigan

Deputy Assistant Secretary of Defense (Requirements and Plans)

About a thousand years ago while I was in college, I took every mathematics course in the catalogue plus a couple more. It seemed like a good idea at the time, but I wasn't really sure why I was doing it, or what the payoff would be. Somehow, however, I knew it was the right thing to do, and it was. Mathematics is a broad foundation on which to build a life and a career. Since the exciting days of college and graduate school, I have done many things, and most of the time my somewhat innocent investment has paid off richly.

My path since school has taken me over terrain I couldn't even have imagined then, but a solid foundation in math has been a pretty reliable compass. I have been a fighter pilot, teacher, analyst, speech writer, and manager. There isn't any prescription for such a career path, but I have always been pretty well prepared.

Flying is one of the loves of my life, and it is all about applied mathematics — time, distance, range, angular rate, and maneuvering in a non-conservative force field. With the proliferation of computers to solve many of the equations of motion which once were the essence of the flying business, one might assume that a working proficiency in math doesn't count anymore — it does. First, you have to verify internally that the solutions on the display panels are about right.

Photograph by Scott Davis, US Army Visual Information Center

Knowing the structure of the problems that are being computed, as well as the technical aspects of the systems, counts. Those who know how things work always seem to work things best. The fundamentals provided by a sound math and technical foundation are the key to flying effectively and safely and, even more important, allow you to apply higher order skills to the problems that occur.

I have moved on to other things. Now I am a Deputy Assistant Director of Defense. I don't use the math that I spent so much time learning directly very often anymore. However, the foundation established years ago is something on which I still rely and upon which I continually build. The primary function of my position is as a long-range planner, and I write the guidance that defines the trade space for resource decisions. A key factor in being able to plan effectively for the future is the ability to estimate trends over time. Many people stream through my door with analyses of a wide variety of problems. With my background, I am able to go beyond the surface of the work and make informed judgments about the utility of the analyses. The instincts and insights that are the product of a mathematics background clearly help in making informed decisions.

Mathematics is a gateway that permits one to take many different paths. I have traveled a lot of them. Whether one wants a career in mathematics or in the broader business world, mathematics is a terrific starting point. If I were starting over again, I would still take all the math courses in the catalogue, but this time I would know why I was doing it.

Richard D. Fuhr

BS Mathematics
University of Wisconsin

MS, PhD Mathematics
University of Washington

Senior Engineer
Electronic Data Systems

Our group at EDS is developing software to represent, manipulate, and analyze various geometric objects. The goal is that the components of our software will be used as a "tool kit" for application programmers who write code for computer-aided design (CAD) programs. These CAD programs will, in turn, be used as tools to design and analyze such things as automobiles.

My mathematical training is applied in several ways in this job. Prior to writing the software, we often need to do research to find or develop the appropriate algorithms. It is not enough that these algorithms be mathematically correct; they should be efficient and numerically stable. My mathematics professors would probably not like to hear this, but I don't really "believe" a theorem or algorithm until I have implemented and tested the corresponding software! That's when it comes to life for me.

Once the software has been written, it has to be documented and debugged, and people must be shown how to use it. I have found my math background to be useful in all of the above tasks. For instance, it is important to provide a clear, concise explanation of what the software does for those who may be fortunate enough to have to maintain it in the future.

Many of the geometric objects that we build are represented as functions made up of many pieces, each of which is a polynomial. These functions are called "splines." Over the last several years an elegant and fascinating theory of splines has been developed, and I have enjoyed learning about this area and applying it to my work. Several years ago I was very active in a national standards committee that adopted a particular form of spline as a standard representation for the exchange of CAD data between different computer systems. This standard is now in widespread use.

Prior to working in the field of CAD, I was in actuarial work. At that time I wrote some notes describing a new technique for solving "stationary population" problems. Recently I was pleased to learn that these notes are still being used 15 years after I wrote them, by people in my old company studying for one of the actuarial exams.

Holly Gaff

BS Mathematics/ Environmental Science
Taylor University
PhD Mathematics
University of Tennessee
Graduate Research Associate
University of Tennessee

When I first started college, I was in a Math Education program, but by the end of my first year, I knew that education was not for me. However, I wasn't sure where to turn. I didn't enjoy math for the beauty of it so I knew that pure math wasn't my calling. I was much more interested in how math could be used to solve real world problems. I started taking biology classes since biology had been my second favorite class in high school after math. In an ecology class, the professor told me about an entirely new area of math that I never knew existed-Mathematical Ecology. After completing my B.S. in Math/Environmental Science, a degree program I created at Taylor University, I found the Mathematical Ecology graduate program at the University of Tennessee and was accepted. I was a teaching assistant for one semester for each of my first four years in graduate school. I have spent the rest of my time in graduate school as a research assistant for a project called ATLSS.

ATLSS stands for Across Trophic Level System Simulation. This project creates a computer simulation of the South Florida Everglades ecosystem. ATLSS is currently part of the Central and Southern Florida Comprehensive Study Review with the goal of aiding plans for major changes to the hydrologic control systems over the next 30 years. This project demonstrates one of the newer areas

of math ecology. It is a multi-model, i.e., a series of models linked together. Each model represents a different trophic level in the Everglades, e.g., hydrology, macroinvertebrates and fish, or particular species, e.g., the endangered Florida panther and Cape Sable Sea Side Sparrow. Each level uses an appropriate type of model, e.g., a coupled ordinary differential equation model, an age/size structured model or an individual-based model. These models are combined to give the best possible simulation of the entire ecosystem.

My specific job is to work on the fish model. I am responsible for writing and running the computer code for the model. I am also responsible for developing tools to visualize the output. The model itself is an age/size structured model and is run over the entire South Florida study area. The area is subdivided into 500m square cells. The fish in each cell grow, reproduce and move both within the cell and among neighboring cells. I use my math skills in both model development and computer programming.

Mathematical ecology and mathematical biology are growing fields. If you have any interest in applied math, I would encourage you to look into these fields. You need a strong background in applied math; in addition, computer programming and some knowledge of biology are very helpful. Six years in graduate school may seem like a long time, but it isn't considering that while I am here I don't just take classes, I gain experience both in teaching and research. When I graduate I can get a job teaching at a university, working for the government or working in industry.

Marshall B. Garrison

BS Engineering
Trinity College

MS Computer Engineering
University of Lowell

MSEE Communications and Signal Processing
Worcester Polytechnic Institute

Principal Software Engineer
PictureTel Corporation

PictureTel is the world leader in the development of video and audio conferencing solutions. During a conference video, audio, and data signals are processed. Two other engineers and I developed a software implementation of a feature known as "continuous presence." This feature combines several video signals into one video signal, allowing conference participants to view multiple persons simultaneously. Another achievement of my department is the 240 Gateway, an internet protocol to standard Public Switched Telephone Network (PSTN) Gateway. The internet and PSTN telephone system coexist at this time. The 240 Gateway permits video terminals conforming to the internet protocol to communicate with video terminals designed to communicate via the PSTN telephone system. I am working on a feature that will enable connectivity between PSTN telephones and Voice over Internet Protocol endpoints. Echoes develop on telephone lines. So, I am installing and tuning an adaptive filter algorithm that will suppress the echoes on the telephone lines.

Notice that the systems I have helped to build are all computer based. The computer science and communications fields have merged, creating a combined computer-communications industry of growing complexity and remarkable capability. There are three prevailing realities.

- There is no fundamental difference between data processing and data communications or between computers and transmission/switching equipment.
- There are no fundamental differences among data, voice, and video communications.
- The lines between single-processor, multi-processor computer, local network, metropolitan network, and long-haul network have blurred.

The communication of information is now the world's largest industry. This reality pushed me back into graduate shcool to pursue an MSEE in communications and signal processing. The courses in my program of study were all heavily mathematics based and assumed an advanced level of mathematical preparedness. I took courses entitled"Principles of Detection and Estimation Theory" and "Digital Communications: Modulation and Coding." The first course discussed techniques for detection of signals in noise and signal parameter esxtimation. The second course introduced various modulation techniques and coding schemes for digital communication over additive white Gaussian noise channels. My prior mathematics education and the prerequisite courses in my program prepared me for success in the two courses. Mathematics is the primary tool for developing, explaining, and understanding the technology of the communications industry.

Ruth Gonzalez

BS and MA Mathematics
University of Texas at Austin

PhD Mathematics
Rice University

Research Specialist
Exxon Production Research Company

As a geophysical mathematician for Exxon, I am involved in the research and development of seismic algorithms that are used in the exploration and production of hydrocarbon reservoirs (oil and gas). An important part of the exploration task is deciphering and integrating various types of data in order to illuminate the subsurface of the earth. Seismic waves propagate and scatter in a complex fashion in accordance with the wave equation, and the seismic reflection method tells us that we can record, process, and interpret reflections from the subsurface geology to produce a picture of reflecting interfaces in the earth. The geometry and geophysical properties which are extracted from the solutions to the wave equation directly impact the drilling decisions for new and existing wells. I develop computer algorithms that use various approximations to the wave equation in order to transform the surface-recorded data so that it is easier to establish the structure of the subsurface.

The accuracy and efficiency of a method are based largely on the velocity variation and geometry that we expect and to what extent they honor the wave equation. For this reason, we need to build a collection of methods so that we can choose the appropriate one for our data. The processing for two-dimensional data gathered over an area with a gently varying velocity field with simple

geology might require only five billion computer operations. On the other hand, three-dimensional data gathered over an area with strong vertical velocity variations would require about 10^{16} computer operations! Even on Exxon's Cray Y-MP supercomputer, these computational requirements, together with the ever-present memory and data-flow issues, make this a formidable and challenging problem.

Real mathematical problems (usually as ordinary and partial differential equations) naturally arise from the physics of wave propagation. It is precisely this connection to the real world that draws me to applied mathematics. My work at Exxon gives me the opportunity to have an impact on the ongoing exploration and production of oil and gas. A prerequisite for producing practical solutions to these mathematical problems is a strong interdependence with other disciplines such as physics, geology, engineering, and computer science. At Exxon I am fortunate to have access to experts in many fields, in addition to state-of-the-art equipment and facilities to conduct my research.

Although the majority of my time is devoted to seismic imaging research activities, I spend time training and consulting with seismic data processors and geophysical interpreters to help them determine which methods are most appropriate for their data. Also, Exxon supports many university research consortia, and I assist in monitoring the research that is carried out within those consortia.

Douglas A. Gray

BS Mathematical Sciences
Loyola College (Maryland)

MS Operations Research
Georgia Institute of Technology

Senior Consultant
American Airlines Decision Technologies Subsidiary, AMR Corporation (Parent company of American Airlines)

Operations research was a natural career choice for me since it combines an interest in business with a strong inclination for the challenge of problem-solving. As an undergraduate mathematical sciences major at Loyola, I was very fortunate in knowing exactly which career path I wanted to follow, and I had supportive faculty members to guide me in the right direction.

At the suggestion of my academic advisors, upon choosing operations research as a career direction, I immediately pursued graduate education as a next step. This turned out to be the best advice I could have received since the MS degree really is the professional degree requirement to do analyst work in the field. Plus, the MS in operations research builds on the problem-solving foundation of an undergraduate degree in mathematics by providing a more in-depth examination of the mathematical tools of operations research, systems analysis and modeling.

Upon completing my Master's degree, I joined the operations research group at American Airlines. The airline industry provides a wealth of complex operational problems. During my first two years at AA I worked in the area of strategic airport planning, analysis and consulting. This work involved employing a sophisticated discrete-event airport/airspace simulation model to evaluate the operational impact of airport capital improvements such as new runways, taxiways, and

terminals. Simulation modeling is a technique commonly used to evaluate the impact of a number of different alternatives in a complex system before committing capital resources to the selection of the "best" alternative. My work took me to a variety of interesting working environments around the world, including Sweden, Spain, and Australia, and introduced me to others in the commercial airline and aviation industry. My strong background in statistical data analysis and systems simulation modeling was most helpful in carrying out these assignments.

At present I am completing the development and implementation of an aircraft overhaul maintenance planning and scheduling system for AA's Maintenance and Engineering Division. This, by far, has been my most successful and rewarding project in that I was completely responsible for the design, development, and implementation of the system. Also, the technical challenge of researching a difficult problem, developing a mathematical model of the process, and implementing the solution in an environment which provides significant benefits to a client is most satisfying. My graduate work in scheduling-optimization models, heuristic algorithms, and linear algebra was most helpful in developing the operations research tools to solve the maintenance scheduling problem both efficiently and effectively.

A wide variety of career opportunities in business and industry abound for those who have the strong analytical, problem-solving skills that are developed while earning a degree in mathematics. These abilities coupled with interpersonal and communication skills, both verbal and written, provide the necessary background to pursue an array of professional opportunities and challenges that are limited only by one's energy, imagination, and ambition!

Cynthia Carter Haddock

BS Mathematics
Missouri Southern State College

MA Statistics
University of Missouri - Columbia

PhD Medical Care Organization and Administration
Cornell University

Professor and Academic Program Director
Health Services Administration
University of Alabama at Birmingham

When I completed my mathematics degree almost 20 years ago, I could not have dreamed that I would be where I am today. Yet, my background in mathematics has prepared me well for what I have done in those intervening years. After leaving college, I completed a master's degree in statistics. While in graduate school, I was a teaching assistant and taught several sections of a probability course each semester. This teaching experience confirmed my love for teaching and the academic environment.

After receiving my master's degree, I took a position as a data analyst with a health planning agency. At this point, the world of health care was very new to me. However, my background in mathematics and statistics enabled me to get the job, since I had acquired analytical and computing skills which the agency needed. Realizing that I enjoyed working in health care I decided to combine my interests in health care and teaching and to pursue an academic career in health administration. As a doctoral student, my major was medical care organi-

zation and administration, and my minors were in organization theory and quantitative methods. Since completing my PhD, I have held faculty positions at Saint Louis University and the University of Alabama at Birmingham.

During 1994–1995 I was on sabbatical, spending the year in Washington, DC, as a Robert Wood Johnson Health Policy Fellow. As a major part of the Fellowship, I worked on health care issues in the Office of the Senate Minority Leader Thomas Daschle.

In 1997, I was named Director of the Residential Master of Science in Health Administration (MSHA) Program in the Department of Health Services Administration at the University of Alabama at Birmingham. This program is a full-time graduate program that trains individuals for executive positions in health services organizations and has been ranked in the "top Ten" nationally by *U.S. News and World Report*. In 1999, I was also named Director of the executive MSHA Program at UAB. This program is a distance education graduate program for experienced, practicing physicians, nurses, other clinicians, and health services managers.

While I continue to teach and do research, much of my time is now devoted to academic administraton. This includes recruiting students, advising, developing curricula, working with faculty, and representing the program to a variety of constituencies. The thinking skills, problem-solving skills, leadership skills, and teaching skills I developed as an undergraduate mathematics major continue to serve me well in these new responsibilities.

Charles R. Hadlock

BA Mathematics
Providence College
MA and PhD Mathematics
University of Illinois
Environmental Consultant
Arthur D. Little, Inc.
Professor of Mathematics
Bentley College

As an undergraduate, I was in an experimental NIH undergraduate program that emphasized a broad scientific background. That's been the key ingredient that let me work my way into many interesting applied problems, and my undergraduate textbooks in physical chemistry, organic chemistry, biology, and physics are still readily available on my shelf. In graduate school I minored in electrical engineering, my first step in trying to do something useful or applied with my math background, but frankly that work was every bit as theoretical as all my math courses. Even though my graduate major was applied mathematics, I felt like a fraud because I had never once really worked on an authentic applied problem.

After several years of college teaching, I decided to see the "real world." Through a contact I had made several years earlier when working on a high school science fair project, I was offered a position at a well-known consulting company whose motto was, "Hardly anything is none of our business." It was true. Here are some examples of the environmental questions I felt privileged to work on.

When and how should our nation dispose of its nuclear waste from power plants, past years of weapons production, and medical and other sources? I've been lowered in buckets into deep mines and flown over large territories in helicopters and planes, trying to identify the pros and cons of different burial or

disposal plans and to develop quantitative estimates of future risks. For example, what's the probability of an earthquake in a certain location, and, if one occurred, could the buried wastes actually start to migrate? I spent most of my time working with geologists, chemists, and engineers.

What is the risk from large industrial facilities or operations like chemical plants, nuclear power plants, hazardous materials transportation, the Alaska oil pipeline? Here we need to develop estimates of both the probabilities and the consequences of all kinds of scenarios that might cause trouble. After doing that, we try to identify the largest contributors to the risk and find ways to counteract them. I spent much time traveling around the world to nuclear plants, similar to the Bhopal plant in India, observing operations, talking to workers, and developing quantitative schemes for analyzing their risks.

How can you clean up a large underground area of contamination caused by many years of leaks from buried chemical drums or tanks (as at Love Canal)? Here we use computer programs based on the solution of differential equations to estimate how the contaminated material is moving today, and how we might be able to get it out by subjecting it to other forces such as artificially changed groundwater flow direction, heat, steam, or air.

There are not enough mathematicians and math graduates going into the environmental field, so the mathematical modeling is usually done by engineers or physicists. The key ingredients are a broad background in basic scientific principles, the ability to learn things quickly on your own, and good communication skills with a wide range of audiences.

Charles Hagwood

BS Mathematics
North Carolina A&T University

PhD Mathematics
University of Michigan

Mathematician
Center for Computing and Applied Mathematics
National Institute of Standards and Technology (NIST)

Because of the many services provided by NIST, it is a multi-disciplinary environment. NIST employs physicists, chemists and engineers of all types, computer scientists, applied mathematicians, and statisticians. I consider NIST a rich and exciting place at which to work. Not only do I have the opportunity to use my education in probability and statistics, but also my undergraduate training in physics and chemistry.

Before joining NIST, I taught mathematics and statistics at the University of Virginia and at Dartmouth College. Now, as a member of the Statistical Engineering Division, my duties are to consult with NIST staff on problems relating to probability and mathematical statistics and to do research on new approaches to such problems.

NIST is a government laboratory with two locations: Gaithersburg, Maryland, and Boulder, Colorado. Its mission is to provide standards for measurements used nationwide and to provide measurement services to support industrial technology. Accurate instruments and products are developed. This process involves several steps of experimentation and analysis. For example, many

measurement processes require the accurate and precise measurement of fundamental constants such as Avogadro's number, the charge of the electron, the proton-electron mass, or the speed of light. Accurate measurement of the speed of light was required, for example, to determine the length of the meter. The meter is currently defined as the length of the path traveled by light in vacuum during a time interval of 1/299,792,458 of a second, and therefore its accuracy is determined by the accuracy of the measurement of the second.

In many cases standardized measurements require the production of certified standard reference materials (SRMs), which are well-characterized, homogeneous, stable materials with specific properties. SRMs are widely used throughout industry and come in a variety of forms. They are used to develop test methods of proven accuracy, to calibrate instruments and measurement systems, to help assure equity in buyer-seller transactions, and to assure long-term reliability and integrity of measurement processes.

One SRM being produced by NIST is blood serum reference material. It consists of a well-blended pool of human serum freeze-dried, then apportioned in equal amounts into 20,000 sealed vials. The purpose is to make available a natural blood serum material that is certified for two dozen clinically important constituents (such as cholesterol, glucose, urea, and sodium) to be used by clinical laboratories to validate their measurements.

John F. Hamilton

BA Mathematics
Cornell University
MA, PhD Mathematics
Indiana University
Research Associate
Eastman Kodak Company

I have always found mathematics interesting and enjoyable. It is also my great fortune to have a job that touches some facet of mathematics virtually every day. I work at the Kodak Research Labs as an industrial mathematician, which means that I am a problem solver. Currently, I am developing new algorithms for Kodak's digital camera program, a task I find both challenging and rewarding. It is satisfying to know that I had a direct hand in developing or improving many Kodak digital cameras on the market today.

Before my current assignment I worked in diverse problem areas such as optics, medical imaging, graphic arts (commercial printing), and laser printers. In each case, I used mathematical models and computational methods to answer a question or solve an engineering problem. I found my mathematics qualifications to be a stepping stone into a whole spectrum of interesting engineering applications.

In industry, the purpose of building mathematical models is to save time and money. In one of my projects, an optical filter had to be designed for a camera. One way to proceed was by trial and error. We could make a filter, test it, and decide how to modify it for the next trial. Each complete trial cycle would take about two weeks, and it could take dozens of trials to get an acceptable filter. Once the mathematical model was developed, the optical performance of any

filter design could be computed in just 15 minutes. Now any number of design ideas could be quickly evaluated with only the most promising actually being made and tested. As a result, we got a better filter and we got it faster.

When I entered graduate school, I thought I would eventually become a faculty member at a university. I was very interested in mathematics and especially liked teaching it. What I found out, as I completed my PhD thesis, was that I much preferred solving "real world" problems rather than proving theorems. In addition, my participation in a summer research project taught me the power of computer simulation. It didn't take me very long to realize that industry held a more promising future for me.

One of the important things one needs to do, especially in industry, is to communicate the results of one's work. Beautiful work that ends up in a file cabinet goes nowhere and just becomes wasted effort. For that reason, the greatest payoff from good mathematical work involves effective communication. But that's just another name for teaching! So, even though I'm a "non-academic" mathematician, a good portion of my time goes into teaching co-workers the details of my work.

William D. Hammers

BS Mathematics
Emporia State University
MS, PhD Mathematics
University of Arkansas
Chief Financial Officer
Optimal Solutions

I am the President of Optimal Solutions, a company I formed in 1995 to provide consulting services for business and education. Located in Clearwater, Kansas, the company helps "organizations make informed strategic decisions and develops software products applying analysis techniques to business problems. The tools of decision support include computers, mathematical modeling and artificial intelligence."

Prior to going with Optimal Solutions, I was a senior program manager in Operations with Cessna Aircraft Company, working with a team to solve problems associated with the manufacture of aircraft. My role was to gather data and extract information from the data for use in making informed decisions. The data mining tools I used included statistical analysis, mathematical modeling, multi-criteria decision support, neural networks, generic algorithms, and fuzzy logic along with the information presentation tools of simulation, risk analysis, and expert systems.

My work at Cessna led me to start my own company to provide decision support consulting services and my mathematics training provided the background I needed. Mathematics plays an important role in the business of manufacturing. Multiple mathematical tools are used daily to model critical business functions which support the manufacturing environment.

Two areas of artificial intelligence, expert systems and neural networks, are particularly important to my work with industry. Expert Systems provide the analyst

with a new class of tools to attack problems whose solution requires intuition, judgment, and logical inferences, as well as facts. Artificial Neural Networks (ANN) also add a different dimension to the analyst's toolbox. Unlike expert systems where knowledge is explicit in the form of rules, ANN generate their own rules by learning from examples. ANN are a multifaceted tool used for speech synthesis, controlling robots, pattern recognition, process control, forecasting and estimating, and many other applications.

Working in industry provides the opportunity to apply mathematics to real world problems and to actually use the results of analysis to resolve difficulties in building a product or delivering a service. I enjoy the challenge of facing problems that are not found in a textbook.

Anyone interested in industry should take courses to improve communication skills, both oral and written. Also take courses that expose you to the current hardware and software tools of analysis. In industry you must be able to explain to business managers how the mathematical tools you have gathered through college courses can be used to solve current problems in real time. And you must also be able to make the correlation in terms that the manager can understand. This requires a familiarity with the current business environment and the tools which are used in it to solve problems.

Carl M. Harris

BS Mathematics and Physics
Queens College
MS and PhD
Polytechnic Institute of Brooklyn
Professor and Operations Researcher
George Mason University

Operations research is the profession that deals with the application of scientific methods to decision making, and especially to the allocation of resources. This work is typically accomplished by developing and manipulating mathematical and computer models of operational systems that are composed of people, machines, and procedures.

Though I have been an academic for most of my professional life, my career has clearly included the major aspects that a trained operations researcher may pursue: industry, academia, consulting, and government. The most exciting part of an OR career is precisely the ability to be able to choose from such a wide range of possibilities. In addition, the very nature of OR as a problem-solving discipline has allowed me to see a wide range of interesting applications, for example, from crime and justice, energy, the environment, manpower planning, manufacturing, telecommunications, and transportation. I have always tried to continue my methodological research in probability modeling and queuing theory, and I have found that to be increasingly valuable in the information technology and telecommunication revolution of our time. All of these interests and activities have combined to give me a very exciting career over the years.

Of the particular applications on which I have worked, the studies that I have found most interesting were those which focused on issues of major public

concern. My study of the Great Lakes and their tributaries brought a colleague and me to brief the U.S. State Department and Environment Canada on our findings. Another colleague and I built a Markov chain model for predicting the spread of AIDS through the turn of the century and its ultimate impact on the health-care system. We have worked closely in this effort with the Virginia Board of Health, the local governments of Northern Virginia, and the U.S. Public Health Service's Centers for Disease Control. An interesting and more recent project got me involved in the government's program to reduce judicial resource requirements for the Federal court system. This work utilized complex queueing models to analyze the flow of cases through the courts and the resultant needs for space and manpower resources.

As an academic, I also do a lot of writing, and my recent efforts have included the co-editing of an encyclopedia of OR and the preparation of a third edition of a textbook on queueing I first co-authored in 1975. A major change in the years since the first edition of this book is the importance now of providing students with computational software and numerous exercises. The development of my own software to do this has been one of the most interesting things I have ever done. It required me to use a very significant portion of the computational mathematics tricks I have learned over the years.

Mary Hesselgrave

BA Mathematics
Newton College of the Sacred Heart

MA Mathematics
University of Wisconsin

MS Computer Science
PhD Formal Semantics
Stevens Institute of Technology

Distinguished Member of Technical Staff
Lucent Technologies Bell Labs

When I was in high school, I knew exactly what I wanted to do with my life—it's just that I was wrong. One of the delightful aspects of a career in mathematics is the opportunity to make choices. I wandered into computer science as part of an effort to be better prepared to teach mathematics for use in today's environment, and I found that computer science is a fertile source of interesting problems. The boundary between theoretical computer science and mathematics is fuzzy, and my dissertation in axiomatic semantics was closely related to formal logic.

I joined AT&T Bell Laboratories after earning my PhD. I have enjoyed working on such different kinds of problems as the development of a semantic interpreter for telephone service orders, performance measurement for an experimental database machine, compiler test automation, performance improvements for database system software, development of a distributed transaction processing facility, operations system capacity planning, systems architecture, and process engineering. The skills involved in my work include digesting large amounts of information to understand the essential requirements for a project. It frequently takes me six months to understand a problem that is initially described in one sentence.

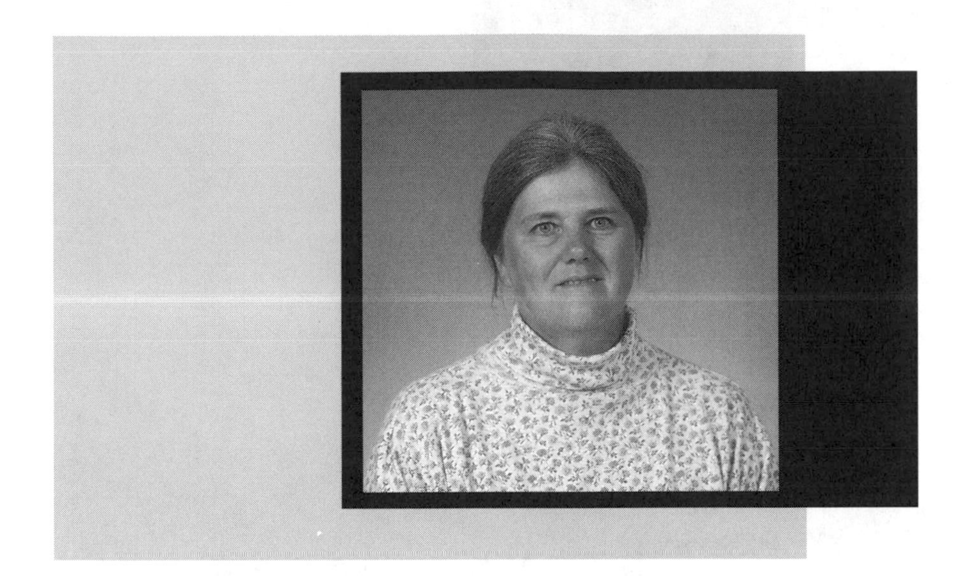

In the course of my job I often work with colleagues to obtain a deeper understanding of the issues involved in our projects. My experience and my theoretical knowledge of computers help identify parts of the system critical to performance; and this experience and knowledge are essential in determining how to measure a system with minimal distortion of the system being measured. I sometimes build or help create a model or validate results from models. From these models and/or results we arrive at the results required to answer customers' questions.

A typical problem in capacity planning may involve determining arrival distributions for different types of traffic, understanding network protocols, and estimating resource use by different types of transactions. One day I may be reading source code, and the next day I may be looking at high-level architecture, so that I am constantly drawing on and increasing my knowledge of computer hardware, operating system internals, compilers, database systems, transaction processing, probability and statistics, numerical analysis, and queuing theory. I like to say that my job involves "learning everything about everything" in the application environment, and it is precisely this opportunity to investigate new areas that brings joy.

I am currently an internal Lucent Technologies consultant on software architecture and performance engineering, working with projects in the United States and Europe. Paying attention to software architecture, performance, and reliability at the beginning of development helps speed new products to market. Things are changing quickly in telecommunications, and there are many interesting problems at the frontier.

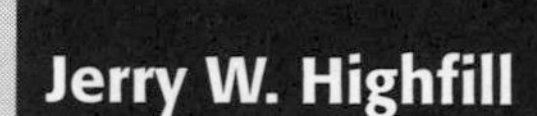

Jerry W. Highfill

BA Mathematics
Southwestern College

MS Statistics
Kansas State University

PhD course work
Colorado State University

Mathematical Statistician
Health Effects Research Laboratory
United States Environmental Protection Laboratory

When higher levels of college mathematics no longer provided practical applications, I began to realize that I wanted no more of that kind of mathematics. I had performed calculations of standard deviations by hand, which almost made me decide to avoid a college statistics course. However, my advisor thought that I should reconsider that decision because the course was simply an introduction to statistics. My advisor was right!

Statistics presents different challenges than those offered by mathematics. Statistics is more wordy, utilizes different logic, and is directed toward a broad range of questions and applications. Yet to comprehend many of the statistical concepts, I needed lots of theoretical mathematics that did not appear to be relevant earlier. Statistics is a great field for men and women.

I have worked as a statistician for three federal government agencies, each committed to health and environmental research. In these professional settings, I have had great latitude in determining which problems to address and how the work should proceed. I have done research for problems arising from water and waste water evaluation and treatment as well as in many fields of animal

testing including pulmonary toxicology and carcinogenesis. Some of the most challenging work has recently arisen from animals exposed to several regimens of ozone at different exposure temperatures where the animals' heart rates and body temperatures were measured every ten minutes for several weeks. I have worked on a study to measure the effect of smoke from coal on humans.

As my knowledge and experience increased, my responsibilities broadened from "helper" to "collaborator" and "principal investigator." Translating a line or two of mathematics into expressive prose at first seemed nearly impossible, but recently it was gratifying to tell research leaders, "This is how we approached the problems of inhalation exposure involving concentration and time and how we approached repeated exposures over several days at several chamber temperatures. Our designs should also function well for compounds. These are the statistical models; you can apply the models to your data if you desire. Would you want to see copies of our publications?"

My mathematics mentor suggested that science and mathematics meet at the highest theoretical levels. This observation still appears to be valid. If you consider working in a technical field, commit yourself early to the study of mathematics and statistics. This knowledge will serve you well, and, with time, it will serve many others.

Harold Jacobs

BA Mathematics
UCLA

MA Liberal Studies
Wesleyan University

Mathematics Teacher
US Grant High School

I have been a high school mathematics teacher since 1962 and have found it to be an immensely rewarding career. Helping students appreciate the value of mathematics both as an intellectual discipline and as a powerful tool for understanding the world is a very satisfying experience. Seeing students become excited about mathematics as they improve their skills and develop confidence in their mathematical ability is thrilling.

Teachers of mathematics, more than any other subject, try to help their students become good problem solvers. The teacher can do what no textbook can do: make the subject come alive by exploring and comparing various strategies that enable students to see that mathematics is an art as well as a science. George Pólya's *How to Solve It* and *Mathematical Discovery* and W.W. Sawyer's *Mathematician's Delight* convey especially well the excitement and intellectual satisfaction that can take place in the mathematics classroom.

One of my primary goals is to teach my students to think. The process of learning mathematics enables one to obtain a better understanding of some of the most profound ways of obtaining knowledge: guessing, generalizing from observations, arguing by analogy, reasoning inductively and deductively. Creating mathematical experiences that enable your students to discover for themselves is a challenge; the satisfaction of success is continually rewarding.

A mathematics teacher's knowledge should not be limited to mathematics. A good liberal arts education makes it possible to draw examples from many subjects into the mathematics classroom. I have also found that the history of mathematics and the philosophy of science have been of immense value. I am continually expanding my knowledge in the process of preparing lessons as well as doing research into areas of my own interest.

The opportunities available to an ambitious teacher are almost unlimited. I know this, not only from my own experience, but also from talking to other teachers. Near the beginning of my own career, I felt the need for a new course. The development of this course led to writing a textbook; the textbook opened the door to other writing opportunities. Creating plans for classroom lessons led to requests to speak at conferences, first local, then state and national. It also led to giving demonstration lessons for the school district and to teaching classes for other teachers at a nearby university. I have found the opportunities to be so numerous that it has become impossible to take advantage of them all. Leadership in one's school or in a mathematics organization, curriculum development, teaching other teachers — there are many needs in mathematics education.

Perhaps some remarks from a letter written to me by a former student say it all: "Your interesting teaching methods made learning fun. I never thought I would understand math, much less enjoy it! I'll always remember you as a teacher who really seemed to care."

Richard Järvinen

BA Mathematics
St. John's University
MAT Mathematics
Vanderbilt University
PhD Mathematics
Syracuse University
Professor and Consultant
Winona State University and NASA

We often hear that mathematics is found everywhere. In a certain sense that is true. There is a need, however, to say that one finds mathematics where one does because someone has put it there. And I believe there is a need to emphasize that mathematics is where someone has put it, in contrast to where one finds it. A successful career in industry depends more on where one puts mathematics than it does on where one finds or observes it.

I have worked as an aerospace scientist with NASA, Remington Rand Univac, and General Electric. I have worked as a medical research statistician with the Mayo Clinic. For most of my professional career, I have been a professor, teaching mathematics, statistics, or computer science. Too often, especially in my industrial employments, I have observed (occasionally frustrated) colleagues waiting to be given mathematics in the form of a well posed problem that will be their work rather than formulating problems that need to be solved and then, themselves, putting mathematics into their work.

I have had the enjoyment and good fortune of creating various mathematical models that have provided explanations and solutions to some important real world problems. Carefully articulating a problem is the most critical aspect in resolving real world problems. Fitting an appropriate model, as vital and as creative an act as that is, is subsequent to deeply understanding what the problem truly entails. Here, fitting a model means writing the perceived problem in an appropriate language, such as the language of mathematics, statistics, or computer science.

Half of my professional hours during the past five years have been given in connection with my role as a Research Scientist at the NASA Johnson Space Center in Houston, Texas. My NASA work has centered on reliability studies and risk assessments for the Space Shuttle program. In 1996 I received from the Director of the NASA Johnson Space Center the NASA certificate for superior accomplishment.

The award was issued for my contributions to the Space Shuttle program. Of timely importance to NASA, I did an analysis of incidents of gas paths at the nozzle-to-case joint on the thrusters of the solid rocket motors of the reusable Space Shuttle. This was for post-Challenger flights. It was a gas path that lead to the Challenger accident in early 1986. Gas path problems re-emerged on the Shuttle thrusters in years following 1986. Using a logistic regression model, I established that the gas path incidents, none of which actually lead to a serious accident, were following a worsening trend. Subsequent to my written and oral

Richard Järvinen and Richard Smally at a summer 1998 Nano Technology Conference held at the NASA Johnson Space Center. Smally had recently won the Nobel Prize in Chemistry.

reports, NASA made adjustments that have reduced the risk of gas path occurrences, and the reliability of the Space Shuttle was increased.

I worked in the aerospace industry early in my career with both General Electric (in satellite detection and in pattern recognition) and Remington Rand Univac (in developing a missile to intercept another missile). I have also worked as a Visiting Scientist in medical research statistics at the Mayo Clinic in Minnesota. At General Electric, while still a graduate student, I developed an optimal searching

Neil Armstrong: Taken by me in summer 1999 on the occasion of the celebration of the anniversary of the first lunar landing.

From left to right: Catherine Coleman, then Vice-President Gore, Eileen Collins. Collins is the first woman to command the shuttle. Gore is complimenting her and the crew on their great work. Coleman launched the Chandra Telescope from the shuttle.

procedure (for maximizing the probability of detecting satellites at all times during a radar search for them) for the Heavy Military Electronics Division of General Electric in Syracuse, New York. That algorithm finds general application as an optimization method in the field of operations research.

Effectively communicating ideas to colleagues in both oral and written form is critical. Here is where I believe my career as professor and my writing experiences have been assets. Teaching requires effective communication. I was honored to receive the Award for Distinguished College or University Teaching of Mathematics from the North Central Section of the Mathematical Association of America in 1997. Writing most definitely requires effective communication. My book, Finite and Infinite Dimensional Linear Spaces, was cited as one of the top ten books written on its subject for permanent library acquisition in a recent publication of the Mathematical Association of America. Skills in both written and oral communication are ever important to develop.

But maintaining and improving physical health also requires consistent attention. I find cross-country skiing and distance running healthful and enjoyable. I have been faithful to a regimen of vigorous physical exercise most days of the year for at least the past 30 years. In February, 2000, I participated for the 25th consecutive year in the 55-kilometer American Birkebeiner cross country ski race, a festive international event held in Wisconsin that attracts some 7000 participants.

Clare Johnson

BS Mathematics
Spring Hill College
MA Mathematics
Duquesne University
PhD Mathematics Education
Columbia University
Professor of Mathematics
Fashion Institute of Technology

I teach mathematics at the Fashion Institute of Technology, which is part of the State University of New York. My students receive Associate (2 year) and Bachelor (4 year) degrees in Art & Design, or in Business & Technology. I prefer teaching students in Art & Design, as the mathematics courses they take are more visual than the skills-based courses the Business & Technology students take. Most of the students at F.I.T. have weak skills and a fear of mathematics. It is fun for me to try to make math accessible to them. They can relate to subjects like the golden ratio, Penrose tiles, knots and mazes, fractals, etc. In analyzing and quantifying shapes, they use logical powers they never knew they had. These logical analyses are used along with their propensity for design.

Teaching Art and Design students at a college for fashion-related studies is very different from teaching at a liberal arts college. To begin with, the students at F.I.T. are dressed in all sorts of imaginative attire, from leather to silk. There are no liberal arts majors. Faculty in Liberal Arts must work hard to find ways to pique students' interest. Teaching these courses the traditional way is guaranteed to lose students' interest.

Under a National Science Foundation grant I developed curriculum for the Art and Design student titled "Geometry and the Art of Design." The 5 major categories in the curriculum are: symmetry, phi and Fibonacci, tiling, polyhedra, and topology. The curriculum is designed to unite the students' creativity with logical processes and to show that mathematics does relate to and can enhance design.

I believe that my artistic background has given me an affinity to this career. For many years I was a dancer and a musician. Currently I play the double bass, playing in community orchestras, in amateur chamber music groups, and with my husband who is a jazz musician.

I gave many lecture-demos at mathematics conferences, showing how music and dance relate to mathematics. My performing arts background has given me a love for relating the arts to mathematics along with helping make me (I hope) a more interesting teacher.

Currently I am chairperson of the Science and Mathematics Department at F.I.T. and do little teaching. This position is more managerial than artistic. Having a math background ensures good problem-solving skills, which is what a manager needs. But I believe that in addition to problem-solving skills, having a variety of interests is of equal importance. I strongly recommend a liberal arts background for any career, including a career in mathematics. No career is uni-faceted. A varied background helps enable a person to deal with the multi-facets of a career.

I am an artist, teacher, and manager. I enjoy my work because I can bring my talents in these three areas to what I do.

Denise Ann Johnston

BA Mathematics
PhD Statistics
Instructor and Consultant on Quality Control
Boeing Commercial Airplane Group

Statistics is one of the most rapidly expanding fields today. Combined with current computer technology, the potential for its future applications is unbounded. The tremendous excitement generated by statistical methods to solve problems and improve quality creates a demand for more people in this profession. In American industry, the improvement of quality is critical to maintaining a competitive position in the world market.

I love my work for many reasons. First, I believe in Dr. W. Edwards Deming's 14 points on continuous quality improvement and its importance to the survival of American industry. Dr. Deming is the man who helped Japan become successful in many industries by combining the use of statistical tools and his 14 points to gain continuous improvement of a product.

Second, I am an avid statistician, and I enjoy applying statistical techniques and teaching them to others. These techniques help us become objective with data rather than allowing us to make decisions based on opinions.

Another great aspect of my job is that it is both investigative and creative in nature. For me, the thrill stems from the wonderful uniqueness of each new problem, with the solution requiring creativity as well as scientific skills. In es-

sence, each application requires me to function as a "quality detective" using process knowledge hand in hand with my statistical skills. The vast array of problems range from manufacturing of parts to computer simulations, from software testing to engineering designs of the process and product. Anywhere a process can be improved, statistical tools are beneficial.

Last, but not least, my job allows me to work with people, to constantly interact with my superiors, my peers, my students, and the public. The advance of technology and the making of a great product happens only because of the hardworking people who create needed changes.

There are many ways to prepare for a career in instruction and/or consulting. My background includes degrees in mathematics, mathematics education, and statistics. However, my colleagues have degrees from many different fields including engineering, physics and chemistry, but most commonly in mathematics. Excellent preparation for this career includes experience in problem-solving using analytical tools, working with teams as a consultant, teaching mathematics or statistics, experience with personal and mainframe computers, and interacting with people. But the most important attribute is a desire for constant, lifelong learning and challenge. In this career, the journey is definitely the reward!

J. Arthur Jones

BA Mathematics and Chemistry
Lincoln University
PhD Mathematics
Pennsylvania State University
President
Futura Technologies

I am President of Futura Technologies, Inc., a company in Reston, Virginia, that I founded in 1989 to provide products and services that apply existing and emerging technologies to enhance the education of school children. These technologies include: instructional video tapes for students, teachers, and parents; and interactive video games that teach as well as entertain and computer software especially designed to motivate and improve traditional classroom instruction.

Initially, the company is focusing on mathematics education in grades kindergarten through 8th grade, but later we plan to expand into other fields and to higher grade levels. I chose this initial focus because K-8 is such a crucial time in the development of basic mathematics skills and mathematical reasoning and because of the potential for technology to improve mathematics instruction. Many young people spend enormous amounts of time playing video games that are designed solely for the purpose of entertainment but have no educational value. My company intends to counter this situation by designing video games and other audio-visual materials that are both entertaining and educational.

During the past several months, I have used my knowledge and experience as a mathematician to create new strategies for improving the teaching and learning of mathematics, including arithmetic, algebra, geometry, probability, and statistics. Some of the strategies involve the "hands-on" use by students of common household items such as cups, egg cartons, and buttons to illustrate mathematical concepts. I am also exploring ways of using interactive videos and computers to simulate hands-on experience through a technique that would allow these devices to display certain objects on command and manipulate them as desired. For students interested in basketball, I have developed materials that use basketball to motivate and clarify several topics in mathematics. This work is an outgrowth of my participation over three summers as an academic coordinator for the Reston Youth Basketball League.

My background in mathematics strengthened by a variety of work experiences in education, government, and industry, has prepared me to undertake the exciting task of forming my own company, Futura Technologies. I am now in a position to follow my creative impulses for adapting technology to teaching and learning mathematics wherever they lead, and I look forward to a productive future.

Patricia H. Jones

BA Mathematics
Meredith College

MAT Mathematics
University of North Carolina

Professor and Chair
Department of Mathematics & Computer Science
Methodist College

As Chair of the Mathematics and Computer Science Department of a small liberal arts college, I am responsible for coordinating all departmental matters. This entails counseling students as well as faculty. I plan all departmental schedules and participate in many campus governing bodies. I also teach a full load of mathematics classes and supervise prospective mathematics teachers in the public schools.

As chair of a department I have many problems to solve. My training in mathematics helps me analyze these problems and organize their solutions. My ability to work hard at a task and not stop until I have finished comes from my years of solving difficult mathematics problems to completion.

I thoroughly enjoy teaching mathematics and working with students and colleagues. By virtue of being in mathematics education, I have had the opportunity to travel abroad. In the fall of 1993, I was invited to attend the first US/Russia Joint Conference on Mathematics Education in Moscow. While the Russian White House was being attacked and burned, I was presenting a paper on how technology is used in our mathematics classes. What an experience! It was interesting to visit Russian schools and meet with Russian colleagues.

Through my experience in Russia, I realized that we face many of the same challenges and share many common goals. However there are differences. US mathematicians recently attempted to provide common criteria for evaluating content and pedagogy of mathematics curricula through three Standards — documents published by the National Council of Teachers of Mathematics (NCTM). In Russia there are no common criteria. In most Russian states, the control of schooling is delegated to more than 15,000 local districts, and there is little attempt to provide guidance. Also, Russia is just beginning to struggle with the use of technology in the classroom, whereas we have given this issue much attention in recent years. In Russia, individual differences have been played down for the common good of the society, but individualism has held sway in the United States. However, educators in Russia are beginning to provide for individual differences, and US educators are encouraging students to work co-operatively and to support each other rather than always competing.

Being a mathematics teacher is very challenging, yet most rewarding. I enjoy working with students as they prepare for their life's work.

Marcia P. Kastner

SB Mathematics
University of Chicago

MS and PhD Applied Mathematics
Harvard University

Research Scientist
Akamai Technologies, Inc.

In the first edition of *101 Careers in Mathematics*, I wrote about my switch from mathematics to applied mathematics, specifically operations research, a field that uses mathematics to model complex systems and optimize their design. I taught courses in operations research at Boston University in the College of Engineering. I later worked at engineering consulting companies applying operations research techniques to real-world problems. My job at the time of the first edition was at MIT Lincoln Laboratory working on algorithms for air traffic control automation systems.

Since that time, I have not only changed jobs twice, but also changed careers. When I explored potential jobs outside the Lab, I kept hitting a job barrier because I had no recent software experience, which most job opportunities required. Although I had done some FORTRAN programming earlier in my career, I had no experience with modern languages such as C, C++, and Java. Fortunately, a friend of mine was working at a start-up software company that offered an intern program. Even though the program was designed for students who were finishing computer science degrees, my friend convinced the company to hire me. I was assigned a mentor, who directed my professional development in software engineering. Six months later, I graduated from intern to permanent staff. For the next two years, I worked

on consulting projects for the government involving client/server architextures. I also developed and taught a training course in distributed object computing with Java. It was a great experience learning new skills and being on the cutting edge of emerging computer technologies.

Recently, I decided to return to working on mathematical problems. By chance, I met an MIT professor of applied mathematics who had just started his own company and was actively hiring. I interviewed for a position and, based on my applied mathematics and software experience, was offered a job. The company, Akamai Technologies, provides an internet content distribution service that uses sophisticated algorithms to make high-volume web sites run faster. The web content of companies that subscribe to Akamai's service flows throughout Akamai's geographically-distributed network of servers, changing routes and locations in response to current demand. When a person requests content from a subscriber web site the request does not go to that company's server. Rather, Akamai's algorithms route the request to the Akamai server deemed "optimal" based on server location and processing load. This reduces congestion at the company's server and consequently speeds up content delivery.

Through this process of changing jobs, I have learned that it is important to keep growing intellectually by working in new and challenging areas. I have also learned that experience in software engineering helps mathematicians expand their career opportunities.

Sherra E. Kerns

BA Physics
Mt. Holyoke College

MA Physics
University of Wisconsin

PhD Physics
University of North Carolina

Professor and Chair
Dept. of Electrical Engineering & Computer Science
Vanderbilt University

I knew little about engineering as a child; most children don't, unless they are related to an engineer. Through my pre-college education, I was exposed to the arts, humanities, mathematics, and the "basic sciences" — I had heard of engineers, but had never met one personally. In college, I met engineering students. Their disciplined thought and empirical perspective made it easy to believe that they were robotic nerds, the anti-matter analog to art and intellect.

It turns out, I wasn't wrong, or right. I was just a normal, purist zealot crossing twenty. I now see engineering through a different scope. Engineers find accurate solutions to real-world, complex problems. This means that they can find and quantify paths through boundaries that cannot readily be crossed by the "purer" sciences. Engineers solve problems as we reach them, punching holes in that mystic fabric of the fundamental quandaries that unravel intuition as they hide understanding. Engineering might be defined as the development of solutions to problems encountered by humans, yet beyond the reach of their comprehensive theories. It is the art and science of solving real problems.

I had a summer job in college, working for a physicist who worked for an engineer. I spent months on a complex derivation. When I presented my results to my boss's boss, he looked only at the bottom line of the last page, shoved the thick solution back toward me, and said, "It's wrong." I retreated, reworked the

entire solution, found a single, computational rather than theoretical, error (which changed the answer by a factor of 10^6). I returned the next day to explain that my equations were correct! Instead of acknowledging my brilliance, he said, "No partial credit if the bridge falls down."

Life changed. The concept of an accurate answer as valuable entered my personal territory as an extension rather than an alternative to profound thought. Still, I preferred the elegance of physics and mathematics to the crass practicality of engineering, until I happened on microelectronics.

Microelectronics concerns the science, design, and manufacture of those itty bitty "chips" inside computers, watches, kitchen and other appliances, and most "electronic" equipment. Inherent in all of this is the intersection of a large number of traditional disciplines to accomplish: The geometric (math) interweaving of 20 or so puzzle-mapped layers of (physics) materials with distinct electrochemical properties comprising (electrical engineering) transistors connected to one another and to energy sources forming (computer science) a system of two-state switches processing (social science) information valued by society in its individuals.

Creativity and diligence in each of these disciplines have together fostered a societal "computerization," a facilitation and expansion of our individual and collective capabilities, a revolution at the edge of global opportunities affecting each touched life.

Gary Kim

BS Mathematics

MS Business Administration
Salisbury State University

Certified Public Accountant

Accounting Manager
Perdue Farms Incorporated

When my family and I emigrated from South Korea to America in April of 1976, I was only 14 years old and entered school as a ninth-grader. As I recall, I didn't speak any English at the time, and I knew that I had a rough road ahead.

Naturally I struggled through classes in history, literature, and others that required heavy reading. However, I always looked forward to my mathematics classes. The reason? Math is a universal language! The same mathematics was taught in South Korea with the same symbols and logic. Wow! Was I glad… I believe that was the key in building my self-confidence in a new environment.

After graduating with a bachelor's degree in mathematics with a minor in computer science, I obtained a job as a computer programmer. Over the next nine years, I designed, developed, and implemented business application systems. During that period, the most important part of programming (other than that it worked) was logic behind the code. There were always many ways to design and code, but there was only one "most efficient" way. And I know that having a background in mathematics helped me analyze complicated requirements and develop clear and concise systems.

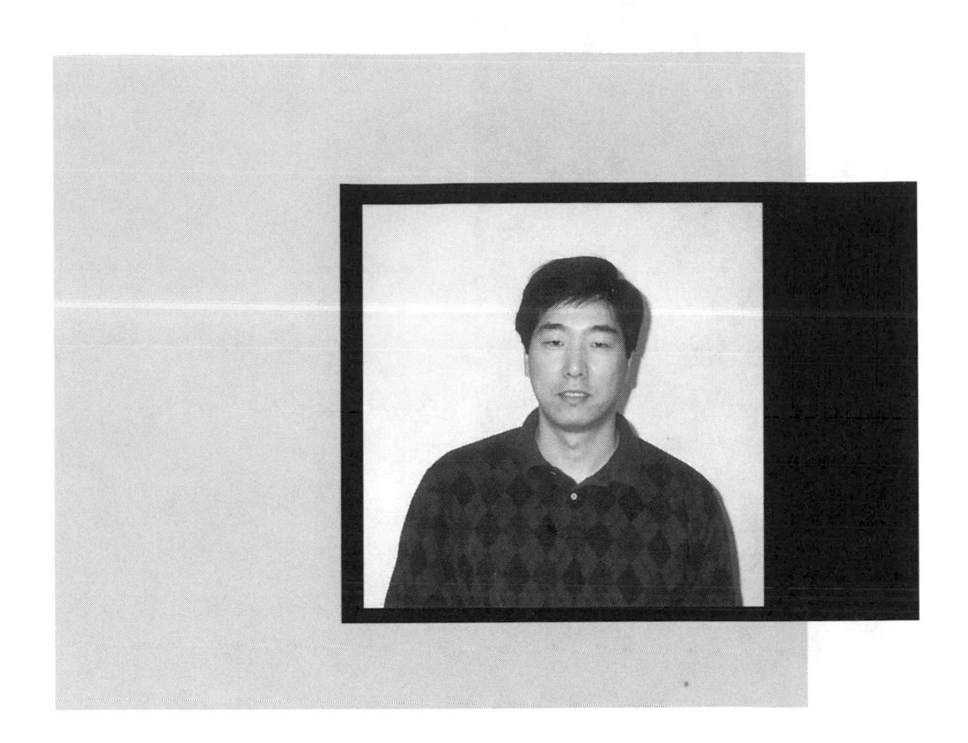

During my study in the master's program, I noticed that mathematics was applied extensively in the business world. For example, determining the optimal capital structure, the break-even point, the point of diminishing return, stock volatility relative to market, and correlation analysis — the list was endless. Again, strong knowledge of calculus, statistics, and algebra was a definite asset.

After receiving my master's degree, I transferred to finance as a financial analyst. This move motivated me to study accounting, which led to my current position. I thought that accounting was basically debits and credits, but I quickly discovered that there is much more. For example, in auditing financial statements, statistical sampling techniques are used to determine the probability of the existence of material misstatements.

Well, it has been almost 19 years in this "beautiful country" (a literal translation of the name "America" from Korean). I think I have come a long way... I also think that one of the critical success factors was mathematics. It was a building block to knowledge. In this changing environment, one must possess the fundamentals of learning — and mathematics certainly is one of them.

Greg King

BS Mathematics
Oberlin College
Teacher
Cincinnati Public Schools

I teach in a magnet middle school with an alternative curriculum. There I have found a wealth of opportunities to use many different parts of my mathemat ics background. My school houses three different seventh and eighth grade programs oriented toward preparing students for college preparatory work in high school. One program is the Cincinnati Academy of Math and Science, which, as the name indicates, focuses on the sciences. Another is a Montessori program, which emphasizes study in many disciplines including all of the arts, and encourages expeditions away from school to learn in non-traditional settings.

The program in which I teach follows the Paideia Model based on Mortimer Adler's "Paideia Proposal." This program emphasizes the development of critical thinking skills in a traditional liberal arts, college-prep curriculum for all students. What makes Paideia unique is it's focus on different modes of teaching. The first is didactic teaching, which is the "normal" lecture-oriented mode found in many schools. The second mode is coaching. In coaching the students are doing the work and the teacher provides guidance. Coaching involves labs in which students are generally practicing and applying previously learned skills or discovering new applications for old ideas. The final mode is Socratic questioning which is done through weekly seminars and range in subject matter through all of the disciplines.

In my program I am a mathematics coach. This means that my classes are math labs, often team-taught with the didactic math teacher. The math labs allow us

the chance to touch on applications in areas of math that many students never see. Students often do lab work in cooperative learning groups. Since our focus is on developing critical thinking skills, labs are designed to be active and engaging. I have had classes work on coming up with a rule for the Four Color Theorem, finding a generalization to figure out how long it would take to solve the Towers of Hanoi, and playing a Fantasy Baseball game.

The third teaching mode, the seminars, are what give Paideia its distinctive feel. The seminar topics rotate through the disciplines, so we have a math seminar roughly once every tour weeks. In the seminars students learn how to form and support an argument from the given text or work and how to argue constructively with each other. The seminars in math have ranged from the children's story "A Grain of Rice" to discussion of the M.C. Escher sketch "Relativity." If you ever want to get students to think about and discuss relative orientation you could do far worse than to use this Escher piece.

While the institutional support for innovative curriculum and pedagogy is more likely to be present in an alternative school than elsewhere, many of the things we get to do I have also done in more traditional settings. One of the most important things I have learned is to challenge students with interesting, relevant problems. Anyone can have students learn about making charts and graphs with a set of numbers written on the board. But, when that learning is folded into a lesson on voting theory or games, you may find that not only are your students more interested and engaged, but you get much more intellectual satisfaction.

Kay Strain King

BA Mathematics
Vanderbilt University

Dip. Ed.; Mathematics, Biology, and Music;
Makerere University College, Kampala, Uganda

MS Mathematics
Texas A&M University

Senior Environmental Mathematician
Theta Engineering, Inc.

What on earth does an environmental mathematician DO? As a member of the Environmental Engineering Division at Theta, a small business which specializes in providing support services to government agencies such as the US Department of Energy and private industries such as Westinghouse Environmental Management Company of Ohio, I usually work as a member of an interdisciplinary team tackling a specific problem on a particular Superfund site. The composition of Theta's team depends on the assignment. The team members must communicate effectively across many academic disciplines and must be able to adequately summarize their work in writing. The work is never boring, occasionally involves travel to interesting places, and has, thus far, used not only every bit of my mathematical and statistical background, but has drawn on all of my other life experiences. After spending approximately 25 years teaching mathematics, I find it fun, exciting, and, I hope, a contribution to our Mother Earth to work with others on environmental assignments.

For example, when Theta did an Environmental Assessment pursuant to the National Environmental Protection Act, our team consisted of an ecologist, an environmental engineer, an electrical engineer, and a hydrogeologist, with me, a

mathematician, as team coordinator. In organizing this project, I used the same logical and quantitative thought processes I used in numerous mathematics courses I have taken or taught.

During another assignment, a Theta team devised a method for predicting measurements of a gas escaping from bermed storage tanks based on weather conditions. We were provided with weather data with readings every hour and gas readings from certain locations also taken every hour — but not necessarily at the same time as the weather data. Because pressure measurements were not available from inside the tanks or on the berms surrounding them, we used dropping atmospheric pressure at the meteorological station to simulate that "the pressure outside the tank is less than the pressure inside the tank." Stepwise regression and time series analyses produced low coefficients of determinations when the independent variables were linear combinations of the meteorological variables. Consequently, we used products of independent variables. We reasoned that, if a temperature inversion did not occur, then, even if the gas were released, it dispersed before reaching the monitors, i.e., it never got measured. Thus the meteorological variables could be split into two categories—those influencing the release of the gas, and those facilitating the measurement of the gas. We ran the final regressions on a contrived (censored) data set and obtained acceptable results.

So, next time you hear a news broadcaster talk about what is happening at your nearest Superfund site, remember that mathematics is an important part of the solution to the pollution!

Maria Klawe

BS and PhD Mathematics
University of Alberta

Dean of Science
University of British Columbia

My life continues to evolve as a kaleidoscope of widely varying, challenging (but fascinating) activities. In 1997, while still continuing as a Vice-President at UBC, I began a five-year term as the NSERC-IBM Chair for Women in Science and Engineering for BC and the Yukon. The goal of this Chair is to increase the participation of women in areas where they are under-represented, especially computer science and information technology. In most other areas of science and engineering the percentage of university students who are female has increased dramatically over the last two decades, but in computer science the numbers have fallen from around 35% to 18%. This makes no sense given the wonderful career opportunities for computer scientists in every field from entertainment and fine arts to education, business, science and medicine.

In 1998, while continuing with my Chair, I started a six-year term as the Dean of Science at UBC. I'm also continuing to lead the E-GEMS project at UBC which studies how computer games can be designed to help children in grades 4-8 like and learn mathematics. Thus my days range from Dean stuff (helping science professors and students with problems about teaching and research) to Chair stuff (working with teachers, students and the media on projects to help girls and women learn computer science and find out about career opportunities) to E-GEMS stuff (working with a team to design and build new computer

games and test them in classrooms). If you're interested in knowing more about my Chair or E-GEMS, you can check out our web sites at http://taz.cs.ubc.ca/swift and http://taz.cs.ubc.ca/egems.

On the home front my husband Nick Pippenger (also a mathematician and computer scientist) and I are having lots of fun watching our two children grow up. Janek is majoring in computer science and plans to be a computer game designer. Sasha is interested in just about everything from psychology to math to languages to sports to music — but not computer programming. In my spare time I train for marathons (I run about one a year), paint watercolor landscapes and play my electric guitar. I have no idea what I'll do after I finish my terms as Chair and Dean. Perhaps be a full-time mathematician or artist. Perhaps write a book or start an internet company and become a multi-millionaire as many of my computer scientist friends have done recently. In any case, I know it will be something I find fun, challenging, rewarding, and worthwhile. It definitely won't be boring!

Karen M. Kneisel

BS Mathematics
Dickinson College

MBA
University of Connecticut

Project Manager
Hydraulic Maintenance
Otis Elevator Company

When embarking on my college career at Dickinson I really did not know what I wanted to be when I "grew up." I knew that I was comfortable with formulas and numbers, but how that translated into a job, I was not sure. When planning my course loads each semester, I simply enrolled in the courses that seemed most interesting to me. Those turned out to be mostly math and computer classes. Thus, I earned a mathematics degree.

After graduating, I wanted to expand my math skills in the business world. I attended graduate school to pursue a Master of Business Administration degree. Math problems had always boiled down to a right or wrong answer for me. However, at UCONN I learned that sometimes there is more than one answer to a problem, and this led me to believe that business and management decisions are very subjective. However, I still did take the classes that were most math-like such as finance, accounting, and statistics.

I went to work for Otis Elevator Company as a production planner after earning my MBA. I was concerned with having plenty of inventory to support the manufacturing operation, but not excess or waste. There is a delicate trade-off between having too much inventory (inventory carrying costs) and having too little

(cost of halting the manufacturing operation due to not enough parts). I applied some Japanese manufacturing principles such as "Just-In-Time" inventory methods. Also noteworthy was the involvement that I had in some process flow analysis on the assembly line which is similar to what an industrial engineer might do. I was eventually promoted to supervisor of inventory control. When I came to Otis the warehouse held over $27 million worth of inventory and now holds less that $18 million. This dramatic reduction is evidence of the success of mathematical theories in the real world.

Although the career path that I chose is not highly technical, it is still challenging and interesting. I have had experience working with people toward a common goal as a team member and as a leader. And as a reward, I have been able to translate my contribution into real dollars saved for the company.

After two years in inventory control, I was promoted again to work with a remote maintenance process that is performed on the elevators and to reduce our costs.

Janice A. Kulle

BS Mathematical Sciences
Loyola College

Senior Product Actuary
The Baltimore Life Insurance Company

Since my graduation from college in 1985, I have worked as an actuary at The Baltimore Life Insurance Company. From the start of my career as an actuarial student to my present position as Senior Product Actuary, I have had the opportunity to combine my interests in mathematics, programming, and business.

As an actuarial student, I sat for tests offered by the Society of Actuaries. Baltimore Life, like many insurance companies, encourages its students to progress through the examinations and often bases promotions and additional work responsibilities on the attainment of the Associate (ASA) and Fellow (FSA) designations. Although the examinations are taken on a self-study basis and at times are difficult, the process is very well structured and extremely worthwhile. Mathematical and business skills as well as a general love of learning are prerequisites for success with these examinations.

Baltimore Life specializes in the development and marketing of life insurance and annuity products. I work with various departments in the company to determine the needs of our customers and design and implement policies that meet those needs. While the growth of computer and software technology has automated much of the day-to-day number crunching often associated with this profession, I often find myself using the logic and reasoning skills which a mathematics background affords.

For instance, two years ago I developed a traditional life insurance product which allows the client to select the number of years of premium payments. I created formulas to generate premiums and then tested their appropriateness in terms of profitability and competitiveness by using a profit testing software package. Without such mathematical formulas, development of this product would have taken much longer. Earlier in my career, when I worked extensively with proposal systems which were written in Basic, I had developed formulas which translated some monthly cash value calculations into annual ones, thereby improving processing time and efficiency. Today, I spend a great deal of time using Lotus, a spreadsheet-based software package. By combining mathematical and programming techniques, I create spreadsheets on a PC which perform calculations that once required mainframe computers to process!

For college students considering a career in actuarial science, I suggest sitting for a Society of Actuaries examination. The first few cover topics such as calculus and probability and statistics, and are a wonderful introduction to the test taking process. In addition, some insurance companies offer internships to individuals who would like to learn more about the profession. Who knows? There may be an actuarial career in *your* future, too!

Kimberly A. Kuntz

BS Mathematics
MS Information Science
Southern Illinois University
Senior Systems Engineer
Electronic Data Systems

As a senior systems engineer for Electronic Data Systems (EDS) I am constantly provided opportunities that require me to apply my *overall* mathematics education.

After completing one year of graduate work in information science, I decided it was time to apply what I had acquired through schooling to the "real" world. I interviewed with EDS during an on-campus career fair and was impressed by their commitment to their policies, philosophies, and value system. Likewise, EDS was interested in my mathematical background and decided that I would be an asset to their company. That was 12 years ago!

In 12 years with EDS, I have been provided numerous opportunities as a programmer, analyst, team leader, project manager, and site manager. I have also had the opportunity to travel throughout the United States providing support to other EDS accounts. I have worked in the health care, life insurance, manufacturing, and special project areas, and I have relied heavily on my overall mathematical knowledge to assist in completing my tasks. There is not one course that I particularly rely on but the mathematical concepts as a whole. As a programmer and analyst, I used my math background to help me think logically through my assignments. As a member of the leadership staff, I used my math background to help develop project plans, staff leveling, estimating and budgeting.

While going to college I had given serious thought to pursuing an education degree but put that on the back burner when hired by EDS. An opportunity to teach was presented to me when EDS was looking for fellow EDSers to become instructors. EDS offers a training program to new employees and looks to those employees who have been out in the field to teach these training classes. Again I relied on my overall mathematics background to assist me in preparing my classes. I taught structured analysis, information processing, and pseudo code to college graduates who had little or no computer science or mathematics background. I needed to start with the basics and then gradually work up to more complex concepts. I reflected back on how I was taught mathematics and how that could apply here, starting out with simple ideas and then moving on to more complex concepts. With the basic approach and with my understanding of the basic concepts of mathematics, I was able to effectively present all the material necessary for the students to grasp the concepts needed to be successful in their next assignment with EDS.

Without my mathematical background I would not have been as successful as I have been in a company whose vision is "to become the premier provider of Information Technology Services on the basis of our contribution to the success of our customers."

Michael H. Kutner

BS Mathematics
Central Connecticut State College

MA Statistics
Virginia Polytechnic Institute and State University

PhD Statistics
Texas A & M University

Chairman, Department of Biostatistics and Epidemiology
The Cleveland Clinic Foundation

Upon completion of my master's degree, I joined the Mathematics Department at the College of William and Mary where I taught statistics, probability, numerical analysis and calculus courses for five years. I loved teaching and applied research and therefore, I went back to graduate school to pursue a doctoral degree in statistics at Texas A & M University in order to get my "union card." My doctoral work included learning statistical techniques and procedures that were especially relevant to the biological sciences and medicine.

I joined the Department of Biometry and Statistics at the Emory University School of Medicine after completing my doctorate. I worked on several interesting studies at the Clinical Research Center. In the cancer area, we showed that providing nutritional supplementation to undernourished advanced colon cancer patients actually worsened their survival outcome. In the surgery area, I was fortunate to work with several surgeons who believed in randomizing patients to determine the best treatment alternative. Here I was able to design and conduct randomized controlled clinical trials to compare surgical outcomes in liver disease.

In addition to working on clinical research studies, I was also publishing methodological research papers in statistical journals. My research interests in linear statistical models afforded me the opportunity to join John Neter and William Wasserman as co-authors on Applied Linear Statistical Models, 2nd edition. This popular textbook is currently in its 4th edition and is continually referred to as the "Bible."

At Emory University, I moved up the academic ranks to Associate Professor and then to Full Professor in roughly ten years. A few years later, I was asked to be Interim Chairman of the Department of Biometry and Statistics. I enjoyed the administrative responsibilities and duties and thus decided to accept the position of Director of Biostatistics in the newly reconfigured Department of Epidemiology and Biostatistics. Over the next two years, Emory University created a School of Public Health. I was asked to serve as both Director of Biostatistics and Associate Dean for Academic Affairs in this newly formed School. In my last two years at Emory, I was solely Associate Dean for Academic Affairs.

In 1994, I left Emory University to join The Cleveland Clinic Foundation as Chairman of the Department of Biostatistics and Epidemiology. The Foundation is a multidisciplinary environment with almost one thousand doctoral level clinical and basic science researchers. The Department of Biostatistics and Epidemiology employs almost ninety individuals. Our mission is to excel in the conduct of clinical and methodologic research. My present work is both challenging and rewarding.

Donna F. Lawson

BA Mathematics and Economics
Denison University

MA Economics
University of Michigan

MS Natural Resources and Environment
University of Michigan

Economist
National Oceanic and Atmospheric Administration

Since July of 1991, I have worked to value natural resources and their services. Why would anyone want to place a price tag on the environment? Well, in the Damage Assessment Center at NOAA we use this information in combination with natural science and legal research to prepare legal claims against parties who harm the environment. NOAA and other natural resource trustees conduct Natural Resource Damage Assessments for oil spills, hazardous waste discharges, boat groundings on coral reefs, and other harmful actions where the law permits us to claim damages on behalf of the public.

Trustees then use the recovered money to restore natural resources to baseline conditions and implement other environmental restoration actions to compensate for resource services that are lost pending recovery. The difficult task of estimating values for natural resource services requires an understanding of economic modeling techniques and of computer programs used for estimation of economic models. In fact, virtually all specialized fields in economics (public finance, international economics, etc.) require a strong foundation in mathematics, with a special emphasis on calculus, probability, and statistics.

In the years since I wrote my original career profile, the old cliché applies: the more things change, the more they stay the same. I still have the same job title (economist), and I still work at NOAA. There have been plenty of changes underlying these simple statements, however. While my job title is the same, I have been promoted in grade, and my responsibilities have changed dramatically — by my own design. After years of working on lawsuits, for NOAA and also in the private sector (anti-trust lawsuits), I needed a change. I still cared very much about the work our office (the Damage Assessment Center) was doing, loved the people I was working with, and felt that I still had a lot to contribute to the damage assessment program, given my experience here. My solution? I moved into a policy role, and have stopped doing case work. I research and work on various issues faced by our program, including tasks such as review of proposed regulations, analyzing information to help us prioritize our program's efforts, and providing peer review on case studies done both within and outside of our program. I am enjoying learning new skills, including the use of geographic information systems software (GIS). I'm working as a 'Jill of all trades' these days, with the variety I need to stay interested and motivated in my work. While the technical details of my math and economics training are not used nearly as often now, I believe that my original training and experience in these areas helped to create opportunities which eventually led to my current position and responsibilities. To learn more about NOAA's Damage Assessment and Restoration Program, please visit our web site at www.darp.noaa.gov.

Christopher M. Legault

BA Biology and Mathematics
Colby College
MS, PhD Marine Biology and Fisheries
University of Miami
Research Fishery Biologist
NOAA Fisheries

Majoring in mathematics does not mean you will be stuck behind a desk grinding out equations. Although I have always had a love of mathematics, this was one of my fears when selecting a college major. Thankfully, I decided to major both in biology, which always fascinated me although I am not the greatest field biologist, and mathematics, which I have always enjoyed and excelled. While looking for a graduate school program to combine my interests, I discovered fishery biology. The field requires an in-depth understanding of the biology of fish and how they are caught as well as the mathematical skills to combine multiple sets of information into a predictive model. This modeling is then used to determine how many fish should be caught next year and the impact of changes in fishing gear. I particularly like the applied aspect of my job - I see the results of my work having an impact on the fish stocks as well as providing opportunities for both commercial and recreational fishing. In addition, I get my feet wet occasionally collecting the data that support the models. Being out at sea hundreds of miles offshore sampling fish around the clock for two weeks certainly is a refreshing break from the office!

Photo of me holding an Atlantic cod (*Gadus morhua*) captured on Georges Bank during the spring 2002 research survey on board the R/V Albatross IV.

All fields of science are becoming more dependent upon mathematical approaches every year. My field, fishery biology, has evolved from simple descriptions of fish life history and rules of thumb to complex models utilizing topics from courses such as Bayesian methods, time series analyses, and bootstrapping. This combination of my interests in biology and mathematics has truly made fishery biology the right career for me. For example, a recent project I completed involved estimating the number of young fish produced by different sized populations of adult fish, known as a stock recruitment relationship. I helped to develop a framework for the analyses that incorporated prior knowledge about the shape of the stock recruitment relationship from other species in a Bayesian statistical framework. A set of 24 possible relationships were examined for each species. These relationships were evaluated against a number of biological criteria and Akaike's Information Criteria was used to select the best fitting model from the biologically plausible ones. This pragmatism of combining statistical rigor with biological understanding is a common occurrence in my field where the data are often quite noisy and incomplete.

Loren Mernoff Lewin

BS and MS Applied Mathematics
SUNY at Stony Brook

Systems Engineer
Bell Communications Research Division

Senior Systems Engineer
Telcordia Technologies

I earned my BS in applied mathematics with a secondary school mathematics teaching certification. I stayed two additional years to earn a Master's degree with a concentration in operations research. After graduation, I was offered a position at AT&T Bell Laboratories as a systems engineer, and upon AT&T's divestiture went to Bell Communications Research Division (Bellcore)—the section of the old Bell Labs owned by the regional telephone companies. In 1997 Bellcore was purchased by SAIC and renamed Telcordia Technologies. Telcordia provides software, engineering, consulting, and training services to optimize the performance of communication networks worldwide, and now has many international clients in addition to retaining the original telephone companies.

As a systems engineer, I have been involved in a wide variety of projects. Due to the varied nature of the work, companies that hire systems engineers look for individuals with well-rounded backgrounds, including proof of the ability to define a problem, analyze it, interact with many others, propose solutions, and communicate them clearly. Thus the wide assortment of applied mathematics, theoretical mathematics, science, and education courses that I took at Stony Brook have all been useful, often indirectly, in allowing me to succeed at my job.

Over the years, my systems engineeting work has continued while my role as a telecommunications consultant has increased. As there is increased competition in the telecommunications industry and the merging of technologies such as telephony, internet, cable, etc., Telcordia is called upon to solve complex technical, engineering, and strategic issues. Our Professional Services organization has an important role in the industry since we are vendor-neutral and understand end-to-end solutions. Though a math major in school and a systems engineer via on-the-job training, I have evolved as the business has evolved into a consultant with all the challenges and rewards of frequent customer interactions. My work has also taken on a greater business focus.

My experience is that majoring in a technical field opened up many varied and interesting job assignments. When selecting a major and than a company to work for, the atmosphere of a company and whether a company can accommodate the lifestyle one wishes to lead will be just as (or more) important than the actual work assignments. I have been fortunate to be able to work part-time since my first child was born, enabling me to have an interesting and rewarding professional life, while at the same time being home with my children after school. I experience great satisfaction from being able to participate in school and community activities due to the flexibility of my hours. Despite working just part-time my contributions to the corporation are recognized to the extent that I was awarded Bellcore's prestigious Distinguished Member of Technical Staff award "for outstanding leadership in the design and specification of …."

Mike Lieber

BS Mathematics/Statistics
Purdue University
MS Statistics
Iowa State University
Biostatistician
Cleveland Clinic Foundation

I'm currently employed by the Cleveland Clinic Foundation, a large, nonprofit hospital, as a Masters-level Biostatistician. I am a member of the Department of Biostatistics and Epidemiology, which—in addition to a large number of computing and administrative support personnel—consists of a number of statisticians (Bachelor's level and up) who are grouped into teams according to the medical specialties of the "investigators" (generally doctors working on research projects) they collaborate with. For example, we have a Cancer team, a Cardio team, a Radiology team, etc. I am part of the Radiology team.

The Radiology team is physically located in the Radiology Department of the Clinic, so there are always patients and doctors passing by my office. I like that; it's a constant reminder that the studies I work on have a direct effect on people's lives, and it helps me keep things in perspective. Being around patients who are here for a CT scan, MRI, ultrasound, or X-ray helps me appreciate the little things in life and minimize the everyday frustrations and stresses we face.

The Radiology team consists of myself and my supervisor/colleague, a PhD statistician. Our work primarily involves helping radiologists with the design and analysis of studies. These studies usually involve assessing the diagnostic ability of the imaging techniques (see previous paragraph) that radiologists use to take pictures of the human body and make diagnoses based (in large part) on these pictures. Often, what is of interest is the "accuracy" of a particular imaging technique with respect to a certain diagnosis. If the accuracy of an imaging tech-

nique, such as an MRI, for diagnosing a torn rotator cuff is not much better than flipping a coin, then MRI is not very useful in making this particular diagnosis. On the other hand, if a doctor is trying to diagnose a rotator cuff as being torn or not, and he or she can be right 95% of the time using MRI, then it's very useful. Studies I've been involved with have included: assessing the accuracy of a new computer-assisted mammography device; comparing the abilities of CT scan and X-ray to detect plastic toys (like LEGOs) swallowed by young children; and comparing the performances of CT scan and ultrasound at detecting kidney stones in male patients.

The statistical analysis of radiologic studies often involves the concepts of sensitivity (avoiding "false negative" tests) and specificity (avoiding "false positive" tests). Diagnostic accuracy is a function of both sensitivity and specificity. Specialized statistical techniques, such as ROC curves, have been developed to address these types of questions.

I consider myself very fortunate to be doing this kind of work. What I do is interesting, I feel that I'm making a contribution, I enjoy working with doctors doing medical research, and the work is neither stressful nor strenuous. I enjoy my job tremendously-more than I would have thought possible when I was an undergraduate at Purdue, wondering what sort of job I could hope to get with a BS in math/stats. It didn't help to be constantly asked, "A math major, huh? So what can you do with that?" Well, the answer to that is good news—there are jobs of all kinds out there for people with a math background.

Peter A. Lindstrom

BS Mathematics
Allegheny College
MA Mathematics
Kent State University
EdD Mathematics Education
SUNY at Buffalo
Retired Professor of Mathematics
North Lake College

What is a two-year college mathematics teacher? Webster's dictionary gives no clues. Is he or she a cross-breed between a high school mathematics teacher and a college mathematics professor? NO!! The two-year college mathematics teacher has very few things in common with these two sets of teachers. The two-year college mathematics teacher is a rare, distinct, and unique breed. He or she:

1) usually has at least a master's degree in mathematics;
2) teaches both developmental math courses (pre-college level mathematics) and introductory college level mathematics courses (freshman and sophomore level); and
3) understands the needs and goals of a typical two-year college student who might be an 18-year-old just out of high school and is not yet prepared for college but is attending because of peer and/or parental pressure; or this two-year college student might be a 35-year-old who is finally getting to college as her children enter high school; or this two-year college student might be a 50-year-old factory worker who, with little formal education, has high goals and ideals with a career change in mind.

The typical two-year college mathematics teacher teaches 15 to 18 credit hours per semester, with four or five preparations that might include an arithmetic course and two intermediate algebra courses (both considered developmental math courses) along with two math-for-business-and-economics courses and a calculus course (both considered college level courses). Also, the two-year college mathematics teacher has the usual out-of-class work: office hours, test and lecture preparation, grading papers and tests, college committee work, and various departmental obligations. All of these activities, both in and out of the classroom, account for more than a typical nine-to-five job.

But these teachers have many other activities that keep them very busy. Many use their spare time either working towards a doctorate in mathematics or studying computer science or some other specialized area of higher education. Others are non-degree students taking courses to fill in deficiencies in their background or to explore further other areas of interest. All two-year college mathematics teachers also have professional obligations to attend meetings and to contribute to mathematical organizations at the local, state, and national levels.

The 27 years that I was a two-year collegemathematics teacher were very enjoyable. I did not become rich, but every day was a challenge with no two days ever the same, and there was much personal satisfaction with the job. There were two other aspects of the job that made it appealing; there were numerous opportunities to contribute to the improvement of undergraduate mathematics education and there was much respect that one receives from students and colleagues.

Tammy M. Lofton

BS Mathematics
Denison University
Actuarial Assistant
Nationwide Insurance Company

Actuaries have been called everything from accountants without personalities to computers that talk. In reality, this profession is exciting with new challenges to tackle every day. People who become actuaries are just as diverse as they are dedicated to finding ways to solve complex insurance problems.

I work in the Commercial Insurance Services department of Nationwide Insurance Co. at the home office in Columbus, Ohio. This department has several divisions that work together to provide technical, educational, and informational support to our regional offices and other departments in the home office. My division is called Pricing/Actuarial. We maintain statistical information on our experience and use it to "predict the future." Using this information, we obtain an idea of how our company is doing and compare this to what is happening in the insurance industry.

The actuarial field is an exciting and growing field. People who take the actuarial exams are well compensated for their efforts and are in demand everywhere. In order to become an actuary, you must pass a series of 10 examinations. They are given once every six months and require a lot of studying. But it's worth it! You gain invaluable knowledge and a great sense of accomplishment. I felt great after learning that I had passed the first two exams. I felt that I had mastered a part of mathematics. Mathematics isn't the only subject that you learn through studying for these exams. Other subjects relating to insurance are covered in later exams. Right now I am studying for an examination that deals with economics and finance.

As preparation for this field, though, it is best to have a strong background in math. The first exam covers calculus and linear algebra — mainly what you would usually learn in undergraduate math courses. The second exam covers probability and statistics, which is another subject that can be learned through undergraduate courses. Other courses in advanced mathematics and in Finance would be helpful as well. Many colleges have actuarial programs that you can follow, but that is not a necessity, I graduated from a liberal arts college with a Bachelor of Science in math.

At first I thought that the actuarial field might be boring. It seemed the only thing actuaries did was count. But, since I loved math, I decided to try it anyway. Since I've been in this field I have learned that my initial perception was totally wrong. It is not just "number-crunching." It's predicting the future and then deciding the best course of action for your customers and for your company. It's working with other people in your company to get the information they need to do their jobs. It's developing new products that your customers need. Also, being an actuary does not limit you to just the insurance field. Many actuaries work for financial and general business consulting firms.

I find personal rewards in the knowledge that I've gained through working in this field — knowledge both of mathematics and of the insurance industry. By choosing this profession, I am a part of the unseen group who help insure a peaceful night's sleep to people worldwide. I'm glad that I chose to join this profession!

Valerie Lopez, A.S.A.

BS Mathematics
University of Texas, Austin

Actuarial Analyst
Towers Perrin

Actuarial Associate
Towers Perrin

It has been 11 years since I started my career as an actuary. I work for Towers Perrin in the area of pension consulting. I obtained my Associateship with the Society of Actuaries in 1995, became an Enrolled Actuary in 1996 and am still pursuing my Fellowship, so I continue to face the daunting task of finishing actuarial exams. I've become involved in the Society of Actuaries Minority Recruiting Committee allowing me the opportunity to volunteer for a cause I believe in within the context of my profession.

Pension consulting involves mathematics from start to finish. After choosing Towers Perrin as consultants, a client entrusts us with establishing and administering its pension plans. We receive client data, and my job is to make sure those data are reasonable and accurate. The reasoning skills taught in math classes that enable you to have a hunch that an answer is right or wrong are precisely the ones that guide me in analyzing a client's data.

I have recently entered the realm of new hire training and management. I still do much of the work listed above, but now I also supervise and train others in this area. I'm now realizing I must especially focus on the art of communicating complex mathematical concepts in a way that can be easily understood.

A successful actuary can not only do the mechanics of their job, but can assess whether or not an outcome is reasonable. In the last few years, I have started to

review other people's work. The review process lets me apply the skills I've learned from going through the mechanics to assess whether or not an outcome is reasonable. For example, if I am reviewing someone's work and a scenario changes the interest rate by 1% and the liabilities change by 50%, but my experience tells me the effect should have only been 10–20%, then I know there may be a problem. I've learned not only the tools to recognize that there may be a problem, but because I know the mechanics, I know where to look for the source of the problem. If something goes wrong, you don't get an actuarial owner's manual telling you what to fix. It's usually years of on-the-job experience that gives you the tools of analysis needed to solve a problem. And because every situation I've encountered is different, I can expect to always improve on the analytical skills I'm developing today.

While mathematics was the foundation that allowed me to start as an actuary, I am learning that a much wider array of skills are crucial to my success. An ability to explain complex ideas in both written and spoken form is necessary in addition to the ability to understand complex mathematics. The skills of a lawyer, accountant, and public speaker are ones that also make a successful actuary. If your strengths are not limited to mathematics, a career in actuarial science might be of interest to you as it has been to me.

Steven W. Louis

BS Mathematics
Denison University
Partner
Andersen Consulting

After graduating in 1978 with a BS in mathematics, I began my career with Andersen Consulting, a worldwide firm which helps organizations reengineer business processes and apply information technology to gain a competitive advantage. Today I am a partner specializing in logistics and materials management techniques, working primarily with clients in the retail/distribution industry. In my work, I am challenged to develop solutions to help companies accelerate the flow of goods and information from manufacturer to distributor to retailer. This often requires the same critical analysis and perseverance necessary to solve a complex mathematics problem. My mathematics experience prepared me to explore and evaluate solutions to problems that involve integrating company strategies with company processes, people, and technologies.

Andersen Consulting is a change agent for its clients. Our business reengineering approach helps our clients to be more successful. In each engagement, I must consider all factors — both internal and external — affecting a company's competitive position. The austere, yet explorative thought process I learned through my mathematics studies, has clearly aided my effectiveness.

At Andersen Consulting, I am regularly required to submit persuasive proposals which outline in great detail how we can help a company become more successful. Clients are obviously more responsive to logical and analytical proposals that clearly explain exactly how a business reengineering approach can help them reap tangible benefits.

Business reengineering is a calculated approach which capitalizes on best practices and often reinvents business processes. As a systems integrator, Andersen

Consulting develops state-of-the-art information systems that support our clients' operations. Each engagement requires precision and critical thinking to successfully implement strategic business solutions.

For example, a manufacturing firm of lighting fixtures wanted to improve the speed and accuracy of communication between central operations and their various satellites. Company sales had doubled over five years, and their current systems—installed in the 1960's—could no longer handle the volume of transactions. As part of the Andersen Consulting team, I analyzed the company's purchasing, manufacturing, and sales operations. From that analysis, we determined how they could eliminate many non-value-added activities and redesign business processes to achieve world-class competitiveness. Our work resulted in companywide changes that saved hundreds of thousands of dollars per year.

My career has provided me with the opportunity to work with clients throughout the United States as well as Europe and Asia. Typically, our firm provides opportunities for people to gain exposure to a variety of client organizations and develop a range of business and technical skills. Andersen Consulting offers challenge, change, and a salient learning environment.

Andersen Consulting recruits new employees almost exclusively from universities. We look for intelligent and highly motivated people who possess the drive to excel and a willingness to work hard. Our clients expect quality work and tangible results. Mathematics provided me with the foundational skills I needed to become a partner with Andersen Consulting.

Christopher M. Luczynski

BA, Mathematics
North Adams State College

MS, Applied Mathematics
University of Vermont

Senior Associate Engineer
Mask House, IBM

I work in the mask house for IBM Microelectronics. That doesn't mean a whole lot without knowing what a mask is, so here is a brief explanation. A mask is much like what a negative is to a photograph. When you want to build a computer chip you create what is called a photomask. It is a glass plate that has the design pattern for one level of a chip on it. The chip manufacturers take the mask and shoot a laser or electron beam through it, and the pattern is etched onto a photoreactive substance. After this is done for several different layers, you have a chip!

My job is as a process/product engineer. Before the chips can actually be built, there is a lot of data manipulation and design work that takes place. Designers send over the data which maps out how to build the chip, but often there are last minute changes or small problems that arise. It is my job to be sure that these changes are still manufacturable and that any problems are worked out before building the mask. This may include visual inspections of chip data, working with customers or just making sure that certain processes were followed correctly. This becomes more fascinating when you consider that most masks deal with patterns that are under one micron thick!

I also do some development work. When new technologies are developing, we often need to make test masks which have certain patterns on them which are

intended to test the abilities of our writing tools. So I get to sit down and build these test masks. This is my favorite part of working in the mask house since I get to do the programming myself and I am immediately able to see the results of my work being built.

I never knew that I wanted to work in a mask house when I was studying math in college. The truth is, I never even knew what a mask was until I interviewed for my job! All through college and graduate school I taught mathematics. As an undergraduate at North Adams I conducted tutorial sessions in trigonometry, calculus and physics. And when I got to graduate school in Vermont I was teaching actual classes in some of those areas while tutoring students in the evening.

I thought for sure I would be a teacher. But for me it was also very important to be able to see the applications of mathematics, not just the numbers. My mathematics background was invaluable to me as training for my career because it shaped the way I think about problems and honed my ability for analytical thought. I may not be cranking out integrals on a regular basis anymore, but the process behind it definitely helped to shape me.

Oh, and I haven't given up teaching. Though I work full time in the mask house, I am also an evening mathematics instructor at Community College of Vermont. There are just some things you don't give up.

Carol J. Mandell

BS Quantitative Analysis
Bentley College

Market Researcher
Corporate Software, Inc.

The next time you watch television or read a magazine, count the number of times you see or hear an advertisement that emphasizes the results of some study or index of customer satisfaction. How many do you think that you would find? I'm sure that in no time you would have a substantial number, but it probably wouldn't exceed the list of research projects that I face every day in my career as a market researcher.

I graduated from Bentley College in Waltham, Massachusetts, with a Bachelor of Science degree in quantitative analysis. I have been with Corporate Software for five years, three as a part-time student and two as a full-time market researcher. Corporate Software is a new company that saw its 1989 sales grow to $92 million. We are in the business of selling PC and Macintosh software add-in products. We sell these products, essentially, by providing a variety of value-added services.

Are these services the ones that best match our customers' needs? Are customers satisfied with the services they receive? These are just two of the questions I research for my company in my job as a market researcher. To that end, I am responsible for the design, execution, analysis, and final presentation of our Quarterly Consumer Satisfaction Survey. I am also responsible for integrat-

ing outside market research services, following the industry through the press and other published studies, and supplying management with the industry information they need to do their jobs.

My interests, both in college and at present, are in market research and, more specifically, in data analysis. The BS in quantitative analysis gives me a distinct advantage over others in the field. That's because, in my work, I need to utilize, understand, and communicate statistics. One purpose of statistics is to draw conclusions about a population based on sample information. Those conclusions determine whether our customers are satisfied and, if not, how we must improve. My degree in quantitative analysis prepared me to do that better than I could with a marketing degree alone.

Market research is one career that you may not see many mathematics majors pursue. It's a career that's fraught with mathematical/statistical challenges, yet rewarding in its power to communicate customer needs. It's a growing field, especially in the area of customer satisfaction. Every day I see more value in my mathematical skills as a market researcher for my company because I am able to interpret customer requirements and transform them into corporate standards of success.

Carla D. Martin

BS Applied Mathematics
Virginia Tech
Consultant
Price Waterhouse

After graduating with a math degree from Virginia Tech in 1995, I wanted a job in which I could use mathematics and have the opportunity to work with many different people. I discovered that the field of consulting was ideal for this because it allowed me to participate in many different projects.

If you have ever heard of Price Waterhouse, you probably associate it with accounting and finance. However, in addition to their accounting and audit service, Price Waterhouse contains a division of consulting services where employees of various backgrounds work on solving complex business problems. I am a consultant with a group known as Management Analytics. This group concentrates on forecasting techniques, market research, statistical analysis, and analyzing large amounts of data. Therefore we seek people with strong quantitative and mathematical backgrounds.

For example, a company was interested in offering a new type of service to its customers. However, before doing so, they needed a way to figure out if this new service would be profitable and determine if the company would gain a large customer base by offering such a service. My first step was to determine the characteristics of potential customers. From that point, I developed mathematical formulas to select a representative sample of US residents whom we could survey to determine their interest in this new service to be offered. I helped design an unbiased questionnaire and supervised the survey adminis-

tration. After we obtained the data, I developed computer programs and formulas to analyze the data. I then wrote a report containing our conclusions and presented it to our client. Other projects I have worked on include one in which we were forecasting future sales of an item for a company. We examined historical company data and built forecast models to determine, based on previous sales patterns, whether the sales will increase or decrease. Our work helps companies plan effectively and raise profits.

But, all formulas aside, my mathematics education has trained me to solve problems logically, and I use that skill in every project I take on. Every day, I use logic to reason precisely, distinguish contradiction from complexity, and determine whether or not a given conclusion has really been proven. Even when no equations are involved, mathematics is an essential part of my job.

I encourage anyone who enjoys problem solving to consider an education in mathematics. You don't have to become a professor to have a financially and intellectually rewarding career in math, and the skills you will develop can be applied in any profession.

Patrick Dale McCray

BS, MS, PhD Mathematics
Illinois Institute of Technology
Systems Project Leader
G.D. Searle & Co., a subsidiary of Monsanto

Searle is a pharmaceutical company engaged in the discovery and devel opment of innovative products for improving health and curing disease. After completing my formal education and a three-year period as an instructor of mathematics at North Park College in Chicago, I applied for a scientific programmer position at Searle. My job at Searle has gone through many changes, but has primarily been focused on the development and sup-port of application software to meet the needs of the scientific activities within research and development. However, there have been periods when I was heavily involved with commercial applications, system software, and administration, and pretty much isolated from the scientific affairs of the company.

The basic process of developing a pharmaceutical is chemistry and biology. It begins with such basic questions as: should it or does it have any activity? Is it safe? Is it effective? The process focuses more heavily on biology, both human and animal. Large amounts of data have to be analyzed at every stage of the process. That's where programming of computers comes into play. We write a program to perform a particular statistical calculation. The statisticians, chemists, and biologists come to programmers, individuals with sufficient background in the sciences to understand the scientific problems and with suitable skills in mathematics and computer science to construct computer programs to solve them.

I consult with statisticians, chemists, and biologists on mathematical questions and on carrying out certain mathematical calculations. I consult with scientists

and software developers on ways to verify whether a specific program is actually doing what it is supposed to be doing: software quality assurance. Recently I led project teams and personally participated in the validation of scientific applications. Currently I am responsible for leading a project to validate a large computer system that keeps track of all the laboratory data at Searle.

When I applied for the job of scientific programmer, computer science was still in its infancy. I got the job at Searle based on my mathematical training and the fact that I had studied FORTRAN programming, had held a technical position at the University of Illinois academic computing center, and had taught beginning programming at North Park College. Today, however, a mathematics major requires a strong background in computer science, the natural sciences, and statistics.

I have been able to provide mathematical consultation by staying alive mathematically. This has been facilitated by being an active member of the Mathematical Association of America and the American Mathematical Society, reading mathematics, solving problems in the journals, and doing major research. Also, over the years, I have been able to assume ever more demanding software assignments by learning from experience, both my own and from others, involving all phases of the system software development life cycle. Reading ACM and IEEE publications and trade journals has helped me keep abreast of developments in computer science.

Jennifer E. McLean

BS Applied Mathematics
University of Colorado at Boulder

Member of Technical Staff
Jet Propulsion Laboratory

What do I do for a living? I send spacecraft to other planets! Okay, I'm not solely responsible for the space program, but it certainly is exciting to be a part of it. Ever since I was a small child, I have been fascinated with outer space. Now I get paid to follow that childhood dream.

I work in the Navigations Systems section at JPL. My duties include the design, development, and implementation of a NASA information system called SPICE for managing geometrical and related ancillary data used to plan space observations and interpret space science instrument data.

For example, when a camera on board a spacecraft sends its data back to earth, the scientists who are studying the images need to know: when was this picture taken? where was the spacecraft at that time? where was the sun? where was the camera pointed? what should it have seen? how fast was it moving? The SPICE software answers these kinds of questions, and it is used in support of several missions including Voyager 2, Magellan, Galileo, Mars Observer, and CRAF (Comet Rendezvous and Asteroid Flyby).

In the applied mathematics program at the University of Colorado, I chose an emphasis in computer science. My computer expertise has proved invaluable

in my work and certainly increased my job opportunities. During my years in school, I worked at IBM doing software integration and testing a large system that operates satellites. Prior to that, I worked at NOAA's Space Environment Laboratory (SEL). The function of SEL is to study solar activity such as flares and coronal holes and their effect on the earth. SEL also seeks to predict when these solar phenomena are likely to occur.

In the summer between my junior and senior years, I participated in the NSF Research Experience for Undergraduates in Mathematics and Computation. I did research in numerical analysis, chaotic dynamical systems, and fractals. I even came up with a result that had never been discovered before.

The goal of this NSF program was to give undergraduates an opportunity to do research which hopefully would inspire them to go on to graduate school. In my case, and in most cases, it worked! I decided back then that I would go to graduate school after gaining some experience in the working world. This fall, after working at JPL for a year and a half, I will begin graduate school in applied mathematics at the University of Washington in Seattle.

Juan C. Meza

BS and MS Electrical Engineering
PhD Mathematical Sciences
Rice University

Numerical Analyst
Sandia National Laboratories

The need for mathematics in industry has increased dramatically in the last 10 years. Nowhere is the need more evident than in the fields of compu tational science and engineering. My particular specialty is known as numerical analysis. It involves a blend of mathematics, computer science, engineering, and physics.

The particular project I have been involved with deals with semiconductor device modeling. The motivation for this project came about from a problem which scientists had noticed in computers that were placed out in space, such as the ones found in communication satellites and space probes like the Voyager. These computers are used to control vital functions such as navigation, and must operate in very harsh environments. Cosmic rays, for example, are continuously damaging the transistors within the computers. Cosmic rays are made up of protons or heavy ions that can have energies exceeding billions of electron volts. There is no easy way to repair a damaged computer once it has been placed out in space, so it is very important that the computers operate for as long as possible to fulfill their mission. I should also point out that there is no known way to shield the computers from cosmic rays, so that the only alternative is to design computers that are resistant to radiation damage.

Can we design a computer which is capable of withstanding this type of radiation? To answer this question we developed a computer model of a cosmic ray striking

a transistor. Roughly speaking, this model can be described in two stages. In the first stage, the cosmic ray travels through the transistor and generates a cloud of electrons as it passes through the material that makes up the transistor (usually silicon). In the second stage, the transistor will react in some undetermined way to this cloud of electrons. In one phenomenon called single-event upset, a transistor which is initially turned off suddenly turns on, or vice versa. If this transistor happens to be located in a vital section of memory, it could cause the computer to fail. One of the jobs of the numerical analyst is to develop the best mathematical methods and algorithms to solve this problem.

Although I have described only my current project, many of the problems which scientists and engineers solve today require the same techniques. As an undergraduate, I studied electrical engineering and computer science, and I learned the importance of a strong mathematical background. After graduation, I worked on various engineering problems, and I began to notice the strong similarities in many of the problems I was working on. I realized that I could work on many problems by concentrating on the underlying mathematical problems, so I decided to pursue a graduate degree in mathematical sciences. Today I get a chance to do exactly that because of my mathematical background. As a mathematician, I have to always be on the lookout for ways to improve the current methods as well as developing entirely new methods. It is never clear what the right answer is, but then that's what makes this job so interesting!

Harlan D. Mills

BS, MS, PhD
Iowa State University

Industrial Consultant
Florida Institute of Technology

We all know that mathematics provides the basis for the physical sciences, the management sciences, and, to some extent, the social sciences. But are we equally aware that mathematics provides foundations for thinking and problem-solving at a more personal level, in ordinary life? Orderly thinking and, just as important, our confidence in orderly thinking help us solve mundane problems as well as exotic ones. I find evidence for this in the unusually large fraction of successful individuals in industry and government with mathematical backgrounds. Often this success comes in jobs that seem to require little math at all.

I have been continually surprised by the level of mathematics education and maturity in successful colleagues in apparently nonmathematical positions or activities. In the Federal Systems Division of IBM in the late 70s I formed two upper-level management committees of the top 15 software executives and managers in the Division. It was a surprise to discover that 14 of the executives had mathematics degrees, several advanced. This was especially surprising because these leaders came from a population of whom less than 25 percent had mathematics degrees. These leaders had been promoted because their projects were successful, not because of their mathematics degrees. Following up on this surprising observation, several people were tracked who had risen, then tailed off. All of these people had soft degrees; none had degrees in hard sci-

ences or mathematics. Before this discovery it had been assumed that what was needed to manage hundreds of programmers in complex projects was simply good interpersonal and communications skills, good organization and planning skills, and much personal discipline and integrity — in short, what any manager of a good-sized organization needs. But from this discovery it became apparent that mathematical education and maturity are more important in managing large software projects than in working on such projects. The management skills are certainly needed, but they are not enough, contrary to what many believed before this study.

In working with these executives and managers, I realized that they were doing mathematics in very general ways. The mathematics they practice is not numerical, not even symbolic, but broader. These leaders drew upon two primary ideas learned from mathematical study: first, the value of structure, and, second, confidence in orderly process.

It is structure that allows people to deal with complex situations in life: to zoom in to the details of trees, then the branches, then the leaves and cells, all the while not losing sight of the forest. The structures of mathematics, for example in algebra and geometry, provide patterns of analysis and synthesis that are of real value in dealing with more general problems, even ones that cannot be easily quantified or symbolized.

Michael G. Monticino

BS Mathematics
University of Florida
PhD Mathematics
University of Miami
Associate Professor
University of North Texas

Without mathematics my life would be considerably less interesting. I have worked on problems ranging from evaluating anti-submarine warfare tactics to forecasting cash inventory needs for a major check cashing company. I spent a summer in South Korea analyzing techniques for detecting North Korean invasion tunnels and another summer working for IBM developing statistical techniques for determining preferences of Internet website visitors.

Entering college, I was fascinated by virtually every subject. I ended up a mathematics major because of a fifteen minute conversation. The chair of the mathematics department at the University of Florida, whose abstract algebra class I was taking, struck up a conversation with me one day after class about my career aspirations. His few minutes of guidance and encouragement literally changed my life.

I received my Ph.D. in mathematics from the University of Miami in 1987. I had every intention of pursuing the standard academic career path. However, I was introduced to the captivating world of applying mathematics to help solve real-world problems during a job interview at an AMS-MAA Joint Mathematics Meeting. As a result of that interview, I accepted a position as an Associate for Daniel H. Wagner Associates, a mathematical consulting company. At Wagner, I worked on problems in anti-submarine warfare, acoustic modeling and optimal ship

routing using mathematical techniques from optimal control, stochastic processes, game theory, search theory, and probability. I also traveled to South Korea as part of a project with the U.S. Army to improve methods for locating invasion and infiltration tunnels that North Korea has constructed under the Korean DMZ into South Korea. My role was to propose more accurate analytical methods for evaluating search strategies. The work involved collaboration with U.S. Army personnel, geologists, electronic sensor experts and South Korean military officers. Working with a diverse team of professionals is typical in applied mathematics. And in this case, I got to relearn some geology—one of the other subjects I was interested in back in college.

I am now an associate professor at the University of North Texas. I am very fortunate to have the opportunity to have a foot in both the academic and the business worlds. I have worked as a consultant for several companies, including IBM, Andersen Consulting, the Federal Emergency Management Agency, America's Cash Express, and the Dallas-Fort Worth Airport. My academic and consulting activities complement one another. The mental discipline, as well as the technical tools, required for research greatly enhance my ability to solve hard problems for clients quickly and within budget. My consulting activities enrich my teaching and often inspire research questions. Moreover, strong communication skills are essential for both effective teaching and consulting. All in all, my mathematical training has allowed me to work on important, challenging problems and continues to enrich my life.

S. Brent Morris

BS Mathematics
Southern Methodist University

AM Mathematics
Duke University

MS Computer Science
Johns Hopkins University

PhD Mathematics
Duke University

Mathematician
National Security Agency

I have one of the greatest math jobs possible: lots of good problems, management that's really interested in their solutions, and plenty of opportunities for professional growth. It's ironic that I didn't want to work at the National Security Agency (NSA) when I started, and I began my career here looking for a "real" math job.

NSA is the largest employer of mathematicians in the US and is part of the "intelligence community," with the mission of solving cryptologic problems. I can't talk about the specifics of much of what I've worked on, but I can tell you something about my career.

After receiving my PhD from Duke, I started in NSA's Cryptologic Mathematician Program (CMP). The CMP is just one of many ways mathematicians are integrated into the NSA work force. Over three years I had tours in five different offices, each giving me a different perspective on NSA's mission. During this period I took NSA courses in cryptanalysis, statistics, number theory, and other advanced math topics. At the same time NSA sent me to school part-time at Johns Hopkins, where I eventually earned an MS in computer science.

Magic and mathematics have always interested me — I even wrote my dissertation on permutation groups generated by card shuffling. During my tours at NSA offices, I learned the perfect shuffle permutation was used to design interconnec-

Photograph by Doug Kapustin© Patuxent Publishng Co.

tion networks. With my interest in magic and my research in the mathematics of card shuffling, I studied these circuits with gusto. The results were several publications and two patents. Most recently the Association of Computing Machinery named me an ACM Lecturer, and I have been speaking at campuses around the country on "Magic Tricks, Card Shuffling, and Dynamic Computer Memories."

During my career with NSA I have had many different jobs. I've worked as a cryptanalyst, where my job was to find patterns in seemingly random sets and to deduce structure where none is apparent. If you don't understand how to break codes, you can't design secure ones, and secure communications are essential to our country's defense.

For four years I taught cryptanalysis and mathematics at NSA's National Cryptologic School. Then, because of my interest in mathematics policy issues, I ran NSA's Mathematical Sciences Program, which awards grants for research in mathematics and cryptanalysis. One of my most challenging jobs was Executive Secretary of NSA's Scientific Advisory Board. This called on all of my technical, diplomatic, and policy skills to help guide a collection of exceptional scientists wrestling with some of NSA's toughest problems.

Today I'm executive of the CMP, running the program I started in. I continue to speak at colleges and universities and now pursue my interest in mathematics policy on several MAA and AMS committees. I think the keys to success in any mathematics job are a love for problem solving, an inquisitive mind, and a willingness to learn new skills. When I run out of new, challenging work at NSA, I'll get back to looking for a "real" math job!

William D. Murphy

BS Chemistry
University of California at Berkeley
MS and PhD Applied Mathematics
New York University

Applied Mathematician
Rockwell International Science Center

After receiving my PhD in applied mathematics from the Courant Institute of Mathematical Sciences (NYU) in 1966, I chose a career in industry rather than academia because I thought that I might have a better chance to observe the development of new areas of applied mathematics, physics, and engineering before the "big guns" at the great universities had a chance to completely investigate and generalize them. I found that I have been able to do this in areas such as computational fluid dynamics, laser optics, device physics, process modeling, acoustics, physical chemistry, and electromagnetics. However, in most cases my major publications have been in physics and engineering journals rather than those in applied mathematics.

My basic task as an applied mathematician has been to model and solve equations in mathematical physics which simulate actual processes, and then let the computer be the experimenter by changing various parameters and studying the effects. In computational fluid dynamics, I solved the transonic potential equation, which is a mixed elliptic-hyperbolic nonlinear partial differential equation. The output is the pressure distribution and lift and drag coefficients for a given airfoil or airplane. The semiconductor device equations are a system of nonlinear mixed parabolic and elliptic partial differential equations. Their output

describes the behavior of the semiconductor as a function of time. In electromagnetics, we are interested in minimizing the scattering from coated conductors to make airplanes "invisible" to radar. This is modeled as a linear Fredholm second kind integral equation with a large number of unknowns ($n >$ 10,000). The magnitude of the separation of the two dominant eigenvalues is related to the power of the laser.

Life as an applied mathematician in industry has not always been a "bed of roses." The years 1969–1970 were particularly devastating years for many of my friends, and I was temporarily converted into a programmer during that time. To survive such downturns in the aerospace industry, it is necessary to develop skills at winning government contracts and solving important corporate problems. This requires the ability to learn new fields of physics and engineering quickly and to "go where the money is."

Knowing both positive and negative sides of life as an industrial mathematician, most of us would have done it all over again in a similar way, perhaps adding a little more physics and computer science to our training in applied mathematics. In the end, one must do what really "turns one on" and not necessarily what is the most lucrative.

My career in mathematics has definitely taken me to heights that I never dreamed were possible when I started at Marist College. My original major was computer science but I realized I was just as good at math as computer science, so I decided to go for a dual major. Four years later I graduated with honors and had earned a free scholarship and Teaching Assistantship at Syracuse University. After two very challenging, and at times difficult years there, I completed my MS in numerical analysis in the summer of 1997. Later that summer I started working for IBM in Fishkill, NY as an assistant Webmaster. I'll admit it helped having a degree in computer science as well as math, but IBM hires a lot of pure math majors, in fact a couple of my co-workers on my current project were also math majors in school.

After a year in Fishkill, I found an opening on the Olympic Internet Team as a Java developer for the Sydney 2000 Olympic Games website in Hawthorne, NY. Using my analytical training, I have to come up with ways to query databases full of statistics from different sports and present them on a web page in a clear and concise way, keeping in mind that not all users are going to be experts in the sport I am coding. The work is difficult, and at times frustrating, but when you finally see your code generating web pages and you see people respond to all the thought you put in, it's worth it. I owe my abilities to my roots in math more than computer science, simply because Java and Web technologies were some-

thing I didn't have a chance to learn in school. They were new technologies and I only got a brief sampling of it by the time I finished my undergraduate studies. I deal with complex problems of data mining, table spacing, and looping constraints more than anything, and, without a solid mathematical base, the most talented of programmers can easily get lost when all these numbers start flying around.

Okay, enough about the hard stuff, let's get down to the reason you're reading this article, the perks !! Working on the Olympic website took me to Sydney, Australia for 2 weeks, and I was able to stop in Munich for Oktoberfest on the way home! I make frequent trips to Madrid, Spain to work with our database team there: I'm seeing parts of the world that I would never get a chance to see otherwise. I work with a lot of talented people, who not only have become my teammates, but also my friends. I really feel a sense of self worth when I hop in my car and drive home at the end of the day or, in some cases, fly home from the far reaches of the earth.

I also have a solid base in teaching math, something I might want to take on as a side job when I have more time My two years of teaching at Syracuse was one of the most rewarding experiences in my life. Nothing beats the feeling you get when students from all different fields of study come to you and tell you math has never seemed more understandable or in some cases even fun with the right teacher. I miss that feeling and hope someday I'll have the time to do it again. As for now, I have more code to write, and plane reservations to make. Good luck and keep working towards your goals, and, if you're lucky, you might find a job that brings you more satisfaction than I get, which is quite immense!

In June of 1990, I graduated from Armstrong State College in Savannah, Georgia with a BS in applied mathematics. I intended to become an actuary. Seeking some real insurance experience, I accepted an entry-level rate analyst position. Working in this field, I discovered I could use my understanding of actuarial information and my love of numbers in a position that offers me a broad range of responsibilities and activities.

Currently, I am a product analyst for AIG Specialty Auto (a division of American International Group, Inc., the largest U.S.-based insurance and non-bank financial services organization). A product analyst's main priority is pricing; and pricing insurance is pricing an intangible. It is pricing a future risk. I must have a solid understanding of the interactions of the marketing, underwriting, accounting and information systems departments. Decisions made in any area can directly impact productivity and therefore impact profitability. It is important that I be able to effectively communicate detailed information with different people both inside and outside the company. It is vital that a product analyst have an analytical and questioning mind to be able to gather information from other areas.

We currently market non-standard auto insurance in twenty-six states and I have responsibility for five. I take great pride in monitoring the results of these states and watching the impact my decisions make on production and profitability. My responsibilities include monitoring productivity, profitability and our competition. Many things impact our product such as the legislative environment, regional economics, case law, and weather patterns. I must be able to understand the impact of a change and implement measures to ensure compliance and contin-

ued profitability. My analytical skills come into play as many changes have a quantitative impact on the bottom line.

Monitoring our competitors' price structure is an extremely detailed project. The market is constantly changing and we must react quickly if we are to maintain or improve our position in the market. A major part of examining our competition includes analyzing their pricing structures. Each competitor can use different rating algorithms, territory definitions, age breaks, discounts, vehicle symbols, violation and accident surcharges, and underwriting guidelines. Using the available information, I must dissect their rates to determine their rating criteria and algorithms. The ability to analyze and understand patterns is very important to analyzing price structures.

I utilize actuarial reports and our competitors' price structures and underwriting guidelines to determine rates and rules that will increase productivity and profitability. Once our pricing structure has been determined it must be presented in a format that can be communicated to all essential parties. The proposal must also be assembled for presentation to the state insurance department. Laws regarding pricing and underwriting practices vary greatly by state. Some states are highly regulated, while others believe in open competition between companies.

As a product analyst, my job is never monotonous, as projects and priorities change almost daily. I work with actuarial exhibits, but much more than "the numbers" help me determine the rate structure I will recommend to build a profitable product.

Joan Peters Ogden

BS Mathematics
University of California, Davis
MS Mathematics
Michigan State University
Consulting Actuary
Joan Ogden Actuaries

I have always viewed myself as a mathematician. As an undergraduate, I took more than half my credits in mathematics and statistics. The rest of my classes were a smattering of one of everything. That breadth has proved very valuable, since as a health care actuary I need to know a little about psychology, sociology, economics, government.... The list is endless.

My job is to predict the future! As a consulting actuary, I am usually called to determine how many people in a given population will develop the need for health care in a future time period (as long as up to 20 years), what kind of care they will receive, and how much that health care will cost.

I own an actuarial consulting firm, and personally consult with four kinds of clients. For insurance companies, I perform the function of an in-house actuary, assist in the conceptualization of health care insurance products, draft policy language, work with claims processing so that they pay claims according to the benefits provided, and research and model to put a price tag on the benefits (premium rate).

For large employers who provide health care benefits for their employees, I evaluate the benefit use, identify areas of misuse or abuse by the employees, and assist the employer in evaluation of and negotiation with its insurance carriers.

For hospital chains and groups of physicians, I assist in determining the right fixed price to charge for packages of services to be marketed to HMOs and other insurance entities.

For regulators and legislators who require information about the future costs and effects of legislation that they are contemplating, I will work on legislative language and evaluation of expected costs and the effect of those costs on behavior. I have been deeply involved in the whole health care reform issue, particularly in my own state.

My predictions, regardless of the client, are based on vast data banks, both those I have developed and others into which I can insert changes in the practice of medicine and the development of new techniques and drugs. I might develop a pricing model that would encompass 20,000 separate calculations for long-term care insurance, or work with a project much more simple, such as the use of a particular pharmaceutical during 1996 among an employed population. My mathematical background is always being called upon for skills in logic, the ability to recognize patterns and anomalies, and the ability to organize large amounts of technical data and to extract needed information. As a consultant, I never know at the beginning of a day what I will have done by the end of the day, but I know it will be interesting and challenging, and will keep me on my toes.

Edna Lee Paisano

BA Sociology and Master of Social Work
University of Washington

Statistician
Bureau of the Census
Department of Commerce

What does the Census Bureau do? The Bureau of the Census is one of the largest statistical agencies in the Federal government. It is in volved in planning, conducting, analyzing, evaluating, and publishing censuses on agriculture, business, construction and housing, foreign trade, manufacturing, population, transportation, and government. Every ten years, the Bureau of the Census conducts the Decennial Census of Population and Housing for delineating congressional districts, determining the number of congressional representatives, and determining the amount of federal funds received for programs.

Our program area develops the data on race groups, including the American Indian, Alaska Native, Asian, Pacific Islander, Black, and White populations. My job is primarily to improve the data for the American Indian and Alaska Native population and their land areas; I am also a manager and a supervisor. I began working at the Census Bureau in June, 1976 and was the first full time American Indian hired by the Bureau. I am Nez Perce and Laguna Pueblo. I have been the key person in developing an inclusive census program for the American Indian and Alaska Native governments and populations. This is very important to me because I am a citizen of the Nez Perce Nation, and the census data we produce are very helpful to tribal governments and organizations in their development of reservation or urban social and economic resources.

What types of projects do I do? We are working on the research and design of the race question we will use for the 2000 census. I am also involved in planning a test census for an American Indian reservation, research on administrative records for tribal governments, strategic planning, writing speeches for the director and other executive staff, conducting workshops and presentations, preparing congressional testimony, working with census advisory committees, designing census reports and computer tape files, and developing outreach and promotion programs. I am the Census Liaison with American Indians and Alaska Natives and Chairperson of the American Indian and Alaska Native Task Force. I have participated in international conferences in Canada, Denmark and Austria.

My education includes a bachelor of arts in sociology and a master of social work. This background may seem odd since I am a statistician, but I grew up in a very rural region where I had limited exposure to math-related occupations. My school counselors never advised me to pursue a mathematics or statistics major , even though I asked them. Because I loved math so much, I studied math and statistics courses in high school and college, and these courses qualified me for my job. I encourage students, especially American Indians and Alaska Natives, to get majors in mathematics, statistics, computer science, or geography because there are only a small number of us in these jobs. These are important fields that have satisfying careers, but they also benefit our people, tribes, and communities.

Bruce A. Powell

BS Mathematics
Denison University

PhD Operations Research
Case Western Reserve University

Manager, Group Control
Otis Elevator Company

The wizard says, "So you want to be a mathematician? Do you know what a mathematician does all day? He works with numbers, computers, and stuff like that." When I was making that difficult decision as to what profession to pursue, that is about all I knew. But that didn't scare me. I liked numbers, computers, and all of the other stuff. Little did I know that I would have such a rewarding career.

The decision of "what I wanted to be when I grew up" was not easy. Having attended a liberal arts college, I was exposed to a number of fields. Believe it or not, during my senior year, I was undecided between architecture, dentistry, the ministry, and good old mathematics. Eventually, mathematics won the battle for my soul.

I began graduate school in pure mathematics, but soon became enamored with the applied side of mathematics that operations research offered. I learned queuing theory in a project with a railroad where we selected the number and location of "pools" of freight cars. I learned about probabilistic modeling applied to inventory theory. I learned about reliability applied to the maintenance of a fleet of repair trucks. I learned about the need for efficient computing applied to sorting a set of numbers. (Back then, computers were slow!)

I spent 20 years at the Westinghouse Research Laboratories working on operations research projects for the operating divisions of the company. My projects covered a very broad range of subjects, and that is what I enjoyed. A short list of fascinating projects that I worked on include (a) nuclear power plant safety, (b) new automated production facility for steam irons, (c) availability of radar units in F-16 fighter planes, (d) design of electric motors, and (e) dispatching of elevators.

"Elevators?" you say. "What does mathematics have to do with elevators?" If you are on the third floor of a tall building and call an elevator in order to leave the building to go home, any one of several elevators might pick you up. The mathematician ... you or me ... must decide the best elevator to be assigned to your call, given that we wish to optimize system efficiency.

I was fortunate to be able to publish some of my elevator research in technical journals. One thing led to another, and I joined Otis at their Engineering Center in 1989. I continue to work in a research mode and have had the good fortune to work in an area in which many of my ideas are turned into patents.

Three points in closing ... first, mathematics is everywhere ... from steam irons to nuclear power plants to elevators. Second, mathematics is hard work. Third, mathematicians are normal people. So, if you like numbers, computers, and all that stuff, go for it! It will be worth the effort.

Marla Prenger

BS Mathematics
University of Dayton

MAS (Masters of Applied Statistics)
The Ohio State University

Associate Scientist
Procter and Gamble

I was one of those rare freshmen who declared a major upon matriculation and never changed it. I chose a degree in mathematics for two reasons: I was good at it, and I enjoyed it. I figured that was all I needed to find a job which was both challenging and fun. Although I wasn't aware of the multitude of career opportunities at the time, I knew that the sound logic skills one hones while obtaining a degree in math would be useful in doing just about anything. How can you go wrong learning skills that can be applied to any type of problem in any job?

At the end of my junior year, I began exploring the option of going to graduate school. While I enjoyed the math classes a great deal, I knew that I didn't want a job doing more or less pure math, and I didn't want to teach. I did, however, want a career working in science as opposed to business. The applied math courses I took gave me a taste of the possibilities of what I could do as a mathematician in the scientific community.

After graduation I went on to get a master's degree in applied statistics from Ohio State University, where my degree in mathematics was obviously a benefit. Students coming into the program with undergraduate degrees in statistics didn't necessarily fare as well as those coming in with a more theoretical background

in mathematics. In any graduate program, experience in dealing with theoretical concepts and being able to logically progress through a complex problem is invaluable. That experience was afforded me by my degree in mathematics.

Currently I work for Procter & Gamble as an Associate Scientist. My primary responsibility is to provide statistical support for various research projects in the arthritis area. As a statistician, my involvement starts at the planning stages of an experiment or assay, where determining an appropriate and efficient study design is imperative to being competitive in the industry. I believe the logical thought processes I developed as a mathematician give me an advantage in drawing on my statistical knowledge to explore and evaluate design options. Solid, critical thinking is required not only to design the study, but also to analyze the data and, in conjunction with the scientists, provide some interpretation of the results. Often the discussion of these results leads to new ideas and hypotheses to be tested.

I thoroughly enjoy the career I've chosen, and I have no question that I wouldn't be here if I had not started my training with a degree in mathematics. The analytical problem-solving skills one develops working through a mathematics curriculum are highly valuable and transferable to any future aspiration.

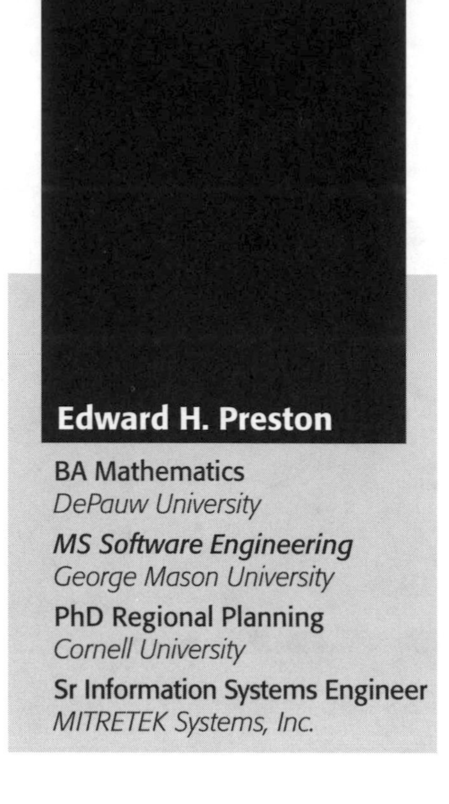

Edward H. Preston

BA Mathematics
DePauw University

MS Software Engineering
George Mason University

PhD Regional Planning
Cornell University

Sr Information Systems Engineer
MITRETEK Systems, Inc.

One of the most rewarding aspects of my job as a systems engineer at MITRETEK Systems is the variety, challenge, and significance of the projects on which I've worked. In part, this is due to MITRETEK's role as a private, non-profit, applied research and engineering organization that is totally dedicated by charter to the public service. We seek solutions to real-world problems of national importance without any product line bias or the constraints of maximizing shareholders' equity. This permits an unrestricted approach to systems analysis and engineering which is both challenging and fulfilling. To meet this challenge, staff members must be able to model and analyze the behavior of a large variety of complex, large-scale, interrelated systems. Mathematics provides an ideal perspective that supports both the required breadth and depth of analysis.

Over the years, I have been involved with many interesting projects dealing with environmental, economic, safety, energy, communication, and information systems. I developed an air pollutant diffusion model and other models to study the environmental impact of various activities. Those models were used to support the establishment of national air quality standards and other regulatory decisions by the Environmental Protection Agency. I built a regional development simulation model and an economic input-output model to study the economic/environmental impacts of activities regulated by the Army Corps of Engineers. I helped develop a series of interrelated, large-scale macro-economic and environmental models to study the effects of policies by the Department of Energy. I helped the US Army develop and integrate their automated command, control, communication, and

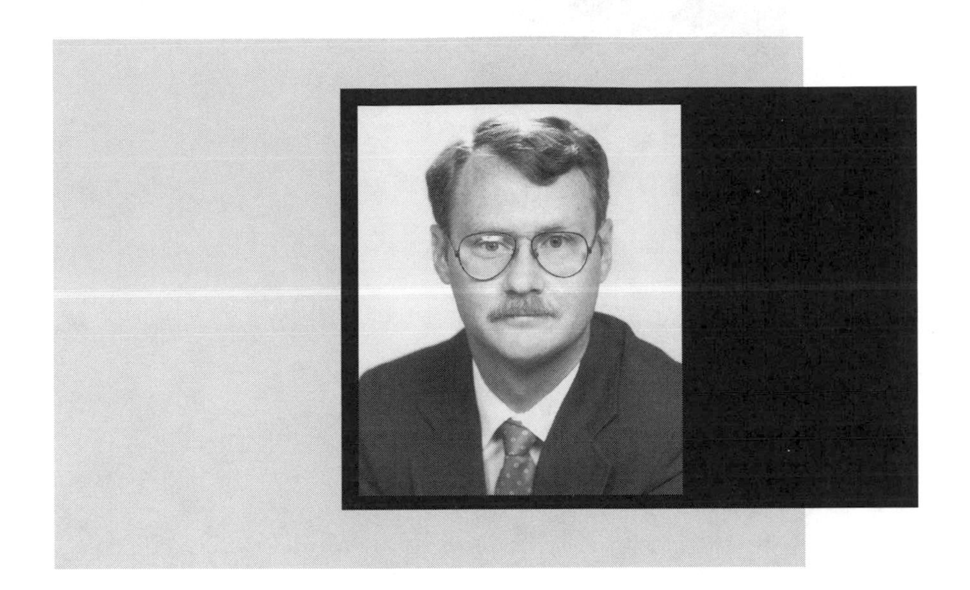

intelligence systems. I contributed to the creation of a NATO-Warsaw Pact wargame using automated planning methods based on artificial intelligence techniques.

Each of the projects I have worked on is different and presents a unique challenge; yet they all require the application of fundamental mathematical concepts and tools. Understanding mathematical principles allows one to see both the differences and the similarities among the various systems and projects. For example, projects analyzing regional economic impacts, battlefield command and control, and IRS tax processing methods all seem quite different. However, once one realizes that each system is basically a network through which "commodities" flow, then common network analysis methods can be applied. The same mathematical techniques can be used whether one is analyzing flows of money, information, or work through a network. It is precisely because real-world projects are not routine, but require creativity and expert judgment to determine how basic analysis techniques can be applied, that a general mathematical background is so useful.

As a systems engineer I have learned that change is inevitable. I have seen improvements in technology, enhancements in analytical methods and tools, and changes in social priorities. Just as evolutionary niches inevitably disappear, so do careers that cannot change with the times. So how does someone prepare for such an unpredictable future? The key is to learn to think, and be able to adapt to new challenges. This is why I believe mathematics is so valuable as an academic background. Solving problems where one is required to extend oneself beyond the cozy confines of rote knowledge is essential in mathematics. It is also essential in order to grow, adapt, and succeed in almost any career in the future.

Fred L. Preston

BA Mathematics
Depauw University

MS Operations Research
Case Western Reserve University

PhD Industrial and Operations Engineering
University of Michigan

Systems Engineer
Lucent Technologies

It is the job of systems engineers to formulate the requirements for new products to meet the evolving needs of customers. These requirements are then followed by hardware designers or software developers in producing the hardware and software. We are also involved in field trials at the end of the development process to ensure that the new product satisfies the customer's needs and is economically viable for the company.

Systems engineering skills can be learned through the study of operations research. Operations researchers use mathematics to help understand and solve real-world decision problems. We need to be good listeners to determine the essence of what the customer wants. We also need to be able to separate out the unimportant details and focus on the critical issues. Often these critical issues can be described and modeled mathematically. Then the whole power of mathematics can be brought to bear to derive solutions. A good systems engineer must be able to see connections between ill-posed problems and potential solution techniques. Sometimes the systems engineer must make design decisions involving the proper setting of parameters that must be "traded off" one for another. These trade-off decisions can be made by mathematically deriving the

best sort of parameter settings or by simulating the system in the computer when the model is complex. Finally, a knowledge of statistics is useful when conducting the field trials of the finished product, testing whether it has met its goals.

My jobs in this field have been interesting and varied. I started working with telephone testing equipment, using mathematics to help determine where companies should place these devices and how many they should purchase. Then I moved to military work. One of the problems I solved was the allocation of anti-missile weapons to attack incoming missiles. Now I am in cellular communications (a field that did not even exist when I went to school). Among my many tasks I use queueing and forecasting models to help cellular service providers predict how much call blocking their customers will experience and when more cellular radios and equipment should be ordered.

Systems engineering is interesting to me because it allows me to be creative. Artists describe their ideas using paint and canvas; musicians use their instruments; I use mathematics. My creations must serve to solve a customer's problem or satisfy an unmet need with a salable product. So I am also a problem solver, and I achieve satisfaction from seeing the results of my creative process helping others.

Systems engineering is a blend of logic and creativity. It is an exciting profession, and it is open to anyone with a logical, creative mind who wishes to use it to solve real problems.

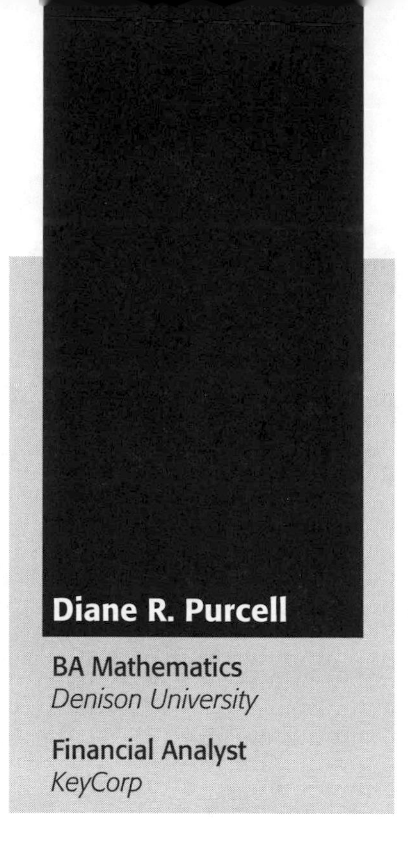

Diane R. Purcell

BA Mathematics
Denison University

Financial Analyst
KeyCorp

Upon my graduation from Denison University in May of 1990, I entered the Management Training Program at Huntington Bancshares Inc., based in Columbus, Ohio. The program was designed to develop numerous skills and abilities in the corporate, retail, and operational sectors of the corporation. The program entailed product and services training in trust, investments, credit analysis, and mortgage lending.

After spending almost two years as a management trainee, I obtained a position as a financial analyst in the Asset/Liability Management Department. I performed monthly forecasts of net interest income and analyses of Huntington's interest rate risk (a measurement of net interest income-at-risk to a directional change in interest rates). In all of my reporting to senior management, accuracy and precision were essential. I produced charts and graphs that were used as a means of communicating asset/liability issues to management. I was also a member of the Retail Deposit Pricing Committee, where I developed reports detailing rates, maturities, and risk characteristics of deposit products. These reports were used in weekly meetings to aid in determining product pricing for corporate-wide markets. For each of these responsibilities, my experience in mathematics allowed me to become more disciplined in my thinking and more attentive to details.

After spending more than two years as an analyst at Huntington, I moved on to a similar position for KeyCorp, a $65 billion bank located in Cleveland, Ohio. I am currently responsible for determining the interest rate risk exposure for affiliate banks in the Rocky Mountain Region. Each day I am faced with numerous responsibilities which require me to use my math and problem solving skills. I formulate strategies and make recommendations to the Chief Financial Officers regarding balance sheet mix and product pricing, and I provide management with a view of the risks inherent in the balance sheet.

In order to enhance my business skills, I entered the evening MBA program at Case Western Reserve University. My undergraduate education in mathematics and my experience in the banking industry will enable me to contribute new ways of thinking to projects that I encounter both in the MBA program and at work.

Experience in mathematics allows for a way of thinking in a structured format. Corporations realize that many types of problems can be solved using the analytical thought process that mathematics requires. A degree in mathematics provides the analytical skills and methods of decision making that are necessary in the workplace today.

Laura Readdy

BS Mathematics/Computer Science
Gonzaga University

Advanced Systems Engineer
Electronic Data Systems Corporation (EDS)

Math was my favorite subject in high school. Looking forward to college, I asked my advisor what I should declare as my major. At the time I thought that majoring in mathematics meant that I would have to teach, but I didn't think that I was cut out to be a teacher. He advised me to major in engineering. After a year, I discovered that engineering was not for me, so I changed my major to mathematics. Still wondering how I could apply mathematics to the real world, I decided to get a combined mathematics/computer science degree.

After graduation I was hired by EDS and have been working for them for more than twelve years. I spent my first four years mainly working with databases. I was a database programmer, designer, administrator, and trainer. I worked on several projects where I worked directly with customers to define and solve their problems. Similarly to solving mathematics problems in school, I had to gather all the facts and come up with the best solution.

After working at EDS for four years, I took a position in their training department. Besides teaching database classes, I also taught a class in testing programs. I found that the theory behind testing programs is all based on mathematics. Of course this was right up my alley. And I had become a teacher!

After leaving the training program, I accepted a position at an account that develops CAD/CAM (computer aided design/computer aided manufacturing) systems. I spent a year training to become a quality advisor. Improving quality deals directly with finding root causes of problems, solving those problems, and using statistics to measure progress. Without my background in mathematics, I would have been lost.

Currently I am working on a project that captures data from the internet and stores it in a database. I'm writing software to populate the database and to retrieve data from the database for reports. This requires a great deal of logic based thinking.

Finally, did you know that mathematics majors make some of the best programmers? EDS hires many mathematics majors and trains them to become programmers. I love what I do, and I know that I made the right decision to stay with mathematics in college.

Brian D. Repp

BS Mathematics
University of Maryland

Senior Data Math Analyst
Bendix Field Engineering Corporation

I have always been interested in the space program and fascinated at human accomplishments in this area. So when I was offered a position at the Goddard Space Flight Center as a Data Math Analyst for the Allied Signal Technical Services Corporation, I didn't have to think about it for very long.

During my first year, I was responsible for analyzing telemetry data from a number of scientific research satellites. Data from these satellites are received and processed at Goddard before being released to various research groups who use the data.

For the next three years, I worked to develop procedures, criteria, and guidelines to be used in the evaluation of data from new satellite missions – specifically the Compton Gamma Ray Observatory (GRO) and, more recently, the Extreme Ultraviolet Explorer (EUVE). I worked closely with the software developers for the computer systems used for data receipt and processing and supported prelaunch testing to verify our group's launch readiness. For the next four years, I worked first as a supervisor for a group of 10 analysts responsible for the receipt and evaluation of telemetry data from the GRO, EUVE and SAMPEX projects, and then as a supervisor for a group responsible for developing a new system in support of the SWAS, SOHO, and XTE missions. Currently, I manage a depart-

ment of 50 people who operate two data capture facilities and provide data analysis, development, and testing support for ten missions.

Although the work I do is not based on any one area of mathematics, I have found that my education has taught me the thought processes necessary to logically approach and work through a problem to arrive at a solution. I use this ability every day to investigate problems by determining under what conditions the problems occur and then testing possible workarounds to be used until the problem can be corrected.

The ability to approach a problem logically and to analyze the possible consequences of different courses of action are the skills I use most in my position. These skills are present in everyone to a certain extent, but tend to be more highly developed in those who have a strong mathematical background.

To supplement my education and further my understanding of the systems with which I work, I am pursuing a degree in computer science. The programming and data structures classes I have taken help me to more fully understand the computer systems I work with and equip me with the ability to better discuss problems and possible solutions with the software developers. My training in mathematics provided me with a great foundation for learning programming.

While education in any field is a worthwhile pursuit, an education in mathematics is one that is used as a basis for many academic career choices.

Mark A. Reynolds

BS Systems Engineering
University of Virginia

MS Operations Research and Management Science
George Mason University

Operations Research Manager
USAir, Inc.

From my senior year in high school through my undergraduate and graduate years, I found it intriguing that the study of mathematics is in many ways like the study of a foreign language. At first, a mathematics student stumbles over fundamental concepts. Over time and with practice, however, these early lessons form the foundation for future study. As one masters the language of mathematics, past approaches are instinctively recalled, just as a fluent speaker in a second language can, without hesitation, negotiate his or her way through an encounter in that language. I do not pretend to be a mathematics master, but I have developed an understanding of mathematics that helps me analyze many of the situations encountered in the airline industry.

Upon entering the University of Virginia, I decided to pursue a degree in systems engineering because of my interest in problem solving. How to put those interests to the best use was not so clear to me at the time. The systems engineering curriculum gave me the opportunity to obtain a broad-based engineering education and develop my skills in human factors, operations research, economics, and computer science. Mathematics was the singular thread throughout all these disciplines. As it turned out, the opportunities upon graduation with my background were both numerous and varied.

My first job, with a federal government contract-research center, provided the means to pursue my interest in operations research. Operations research seeks to use mathematical algorithms to allocate scarce resources. The interesting part of this field is formulating mathematical models that describe meaningful problems, working out solutions of the models, and then interpreting the results to those who posed the problem. During my graduate program, it became clear that the airline industry offered many interesting operations research problems: scheduling pilot crews, assigning aircraft, planning manpower requirements, and managing the flight yields. At Northwest Airlines and currently at USAir, I have had the opportunity to define problem approaches, model them mathematically, develop solutions and guide the implementation of the results. The successful conclusion to these projects saves the company millions of dollars.

As I gain more experience with USAir, mathematics continues to be the central focus of all that I do. Being able to translate business situations and problems into a mathematical model is an invaluable skill to possess in the corporate world. However, the full mathematical contribution can be achieved only by using "people" skills to obtain a clear understanding of the problem from the customer and then to interpret the "numbers" solution.

Nancy K. Roderer

BS Mathematics
University of Dayton

MLS Library and Information Science
University of Maryland

Director
Harvey Cushing/John Hay Whitney Medical Library Yale University

I majored in mathematics and computer science as an undergraduate because I found them fascinating ways to explore the world of logic; I pictured a future in the ivory towers of academe, thinking about abstract number systems and adding to the body of knowledge in the area. I was diverted from this path by another interest, the more practical world of information systems. There, I found, I could apply the general approaches to problem solving that I had learned in mathematics to real world situations and improve people's access to the information that they need.

My fascination with the way in which people find information began before I started school, at my local public library. I was intrigued by the vast quantity of information there, and sensed that it could give great power to those who used it. As I went through school, I found myself increasingly aware of all of the tools – directories, catalogs, indexes, and databases – that existed to help people find information. My first professional jobs after I received my mathematics degree were as a computer programmer, and there I found myself interested in the question of how the programs that were written could best be shared with others. I also began to see how the power of computers could be used to retrieve information rapidly and efficiently.

Photography by Joseph

Since I completed my library and information science degree I have been fortunate to work in a variety of positions where I could continue to study how information is used and how information systems can be improved. I worked as a consultant to library and information services for a number of years, and have spent the last eight years working in medical libraries. Here I have worked directly with physicians and other health care providers and learned about their particular needs for patient information, administrative data, and other resources. The library works closely with other information providers, and together we assure that the needed information is at hand when and where it is needed.

My mathematics background has contributed to my work in both general and specific ways. Generally, my coursework taught me how to identify a problem to be solved, to specify the environment in which that problem exists, and to attack the problem in a systematic way. Statistics courses have helped me in my efforts to survey users and evaluate information system use. And finally, the courses that I took in computer science as a part of both my BS and MLS gave me the basis for much work in developing new, computer-based information services.

Michael Rosenfeld

BS Mechanical Engineering
University of Michigan

ME Mechanical Engineering
Carnegie-Mellon University

Senior Structural Engineer
Kiefner & Associates, Inc.

Many people are unaware that almost all of the crude oil, aircraft fuel, home heating oil, and natural gas consumed in the US is transported by pipeline, sometimes over thousands of miles, on the way to its final customer. Commodities such as coal slurry, propane, fertilizer, and industrial chemicals are also transported by pipeline. Some pipelines are over four feet in diameter and operate at very high pressures; others are quite small. They cross rivers, swamps, mountain ranges, deserts, farms, suburbs, and cities. Some pipelines date back to the turn of the century and are still in operation. The company I work for, Kiefner & Associates, provides engineering services to help keep pipelines such as these operating safely, reliably, and economically. My job challenges me to devise practical solutions to a variety of difficulties encountered in the field, such as controlling the stresses in the pipe due to ground movements caused by underground mining, analyzing vibrations in a pipeline induced by flood waters, or figuring out what caused a pipeline to fail. Mathematics is a valuable tool for understanding our clients' problems and for developing the best possible solutions.

I was trained as a mechanical engineer, not a mathematician. In fact, mathematics was not of great interest to me in college at first. But I quickly recognized that engineering cannot be learned or practiced properly at any level without mathematics, because mathematics is so fundamental to understanding the behavior of structures, machinery, materials, and processes, and also to writing

and using computer programs. As I studied more mathematics, my understanding of science and engineering deepened.

My first two jobs provided me with a wide variety of engineering experiences: simulating earthquake loads in nuclear power plants, evaluating structural integrity of aircraft components, designing electric motor housings to use the minimum amount of material (which earned me a patent), experimenting with steel-cutting operations in flammable atmospheres (which earned me the nickname of "Boomer"), and analyzing equipment failure. These problems all required the modeling of physical phenomena mathematically. Special computer programs were used occasionally to analyze the more complex problems. A mathematical understanding of the physical problem was often necessary to detect whether the computer models were in error and why. In fact, the use of analytical computer programs makes a good grounding in mathematics as necessary as ever.

I have spent most of the past 15 years doing analytical engineering work. This led me to my present work opportunity. Kiefner & Associates is a small company — only four engineers and a couple of support people. Our clients consider us to be the experts and trust us to give them the answers they need. I find this exciting because we must be highly responsive to our clients' needs while having multiple responsibilities in our company. A good solution could save our clients a lot of money and lead to more business, but errors could prove costly or even dangerous. Mathematics has been central to these activities, and it has helped us gain a reputation for quickly developing workable solutions to our clients' problems.

Peter Rosenthal

BS Mathematics
Queens College of City University of NY
MA, PhD Mathematics
University of Michigan
Professor
University of Toronto

From childhood to the present, I have always found mathematics to be the most beautiful of subjects: the elegance of the nicest proofs impresses me more than any other art. This is the reason I studied and continue to study mathematics, although I now also have a second career.

In 1969 I was a young Assistant Professor of Mathematics at the University of Toronto. The war in Vietnam was the cause of frequent protest demonstrations in front of the US Consulate. I was giving a speech at one such when the police told me to stop. I kept speaking. The head of the riot squad told me he would arrest me if I continued. I did, and he did. I was charged with two minor but criminal offences.

Upon reflection I realized that it would be a drag to have a criminal record, so I worked hard preparing my trial with a lawyer. When my lawyer and I disagreed about trial tactics, I fired him and represented myself (with the judge emphasizing that I had a fool for a client). I was acquitted of one charge and convicted of the other at trial; on appeal I was acquitted of the second charge as well.

In the course of representing myself I learned a bit about criminal law and procedure. Over the next twenty years I represented many demonstrators for various good causes. I was a paralegal, with no training but lots of enthusiasm. On several occasions I was frustrated by my lack of a degree in law.

Therefore, in middle age, I decided to go to law school. It was interesting becoming a student again. Anyway, I got through law school (and even enjoyed it a bit) while continuing mathematical teaching and research, and was called to the bar of Ontario in 1992. Most of my cases are still representing protestors of one kind or another, and most of the cases cost rather than earn money. But I get a lot of satisfaction from contributing in that way to causes I believe in.

I still do research (in the theory of operators on Hilbert space) and teach mathematics. My legal and mathematical lives are generally as distinct as Dr. Jekyll's and Mr. Hyde's.

There have been a few times when being a mathematician helped me with law. On one occasion, my being a mathematician gave me the courage to really probe an "expert" witness who claimed that the "flexion" of a graph proved my client was guilty. Although the expert claimed that the definition of "flexion" was too complicated for the court to understand, cross-examination revealed that the "expert" had no real definition in mind. We won the case. Such instances are rare, however. Generally, mathematics and law are very separate (sometimes competing) parts of my life. They each provide great satisfaction, and some tension and unpleasantness. I'm glad I make a living as a math professor: I would not enjoy having to take legal cases I didn't believe in just because I needed the money. Fermat, on the other hand, made a living as a lawyer and did mathematics as an amateur. I know that his mathematics was much better than mine; I wonder if I'm as good a lawyer as he was?

David S. Ross

BA Mathematics
Columbia University
MS, PhD Mathematics
New York University
Research Associate
Eastman Kodak Company

The Computational Science Laboratory at Eastman Kodak's research labs is a group of mathematicians and mathematically-inclined scientists and en gineers who work as internal consultants. I have spent my entire career at Kodak-11 years now-as a member of this group. When a new research project is begun, if mathematical modeling will be required, a member of our lab is asked to sign on. Such projects often last for years. We also work on shorter projects in collaboration with other engineers and chemists. For example, if an engineer wants to understand how some fluid waves affect a production process, we might spend a month or two developing a model of the phenomenon. Many of us have standing collaborations with other scientists in the labs, experts in some field, and we work regularly with them on problems in their fields. Also, we are known throughout the labs as experts in mathematics, so people often drop by our offices for help with short mathematical tasks-to solve some ODE's, to help formulate a system of algebraic equations, to perform a regression analysis, to give our opinions on the best software for a particular application, to find the roots of an equation, to compute eigenvalues, etc.

My specialty is differential equations. These days, I am working on fluid dynamics problems.

One of these is the problem of dynamic surface tension in fluid curtains. The chemically active part of photographic film consists of silver halide crystals and

dye-forming chemicals in gelatin. This is coated on a hard backing in liquid form —it's liquid gelatin with stuff suspended in it, just like a strawberry gelatin dessert with cling peaches suspended in it. The thin liquid layer is then dried. One method we use to coat this liquid is curtain coating, in which a thin, controlled waterfall of the gelatin solution coats the hard backing as it runs under the waterfall on a conveyor belt. Surface tension tends to break up such curtains into drops and rivulets in the same way that water from a faucet tends to break up into drops. We put surfactants, chemicals that reduce surface tension, into the fluid to prevent the curtain's breaking up. In collaboration with some surface chemists, I have developed a mathematical model of the transport of surfactant in coating flows, its diffusion to surfaces, and its influence on surface tension. Mathematically, the model takes the form of a system of reaction-diffusion equations. These are equations like the heat equation, but with nonlinear source terms.

Another project on which I am working is ink jet printing. Because I knew about fluid surface tension from my work on coating flows, I was asked to model drop formation and ejection from ink jet printers. This problem involves the solution the Navier-Stokes equations with free surfaces with surface tension.

In my years at Kodak I've worked on many, many different projects involving many different types of mathematics, from lens design and solid state physics through crystallization and the chemistry of photographic development to statistical analysis of color balance data from photofinishing shops.

Edward Rothstein

BA Intensive Mathematics
Yale University

MA English Literature
Columbia University

PhD Committee on Social Thought
University of Chicago

Cultural Critic-At-Large
New York Times

My career has been such a strange one so far, I am not sure whether I can suggest that any math student try to follow in my footsteps. At first glance it can almost seem, topologically speaking, discontinuous.

I majored in intensive mathematics at Yale University and completed my course work for my MA in mathematics at Brandeis University, but after that I left the field of mathematics completely.

I started writing about a variety of subjects for major intellectual publications like *Commentary*, *The New York Review of Books*, and *The American Scholar* about literature, music, and politics. I went on to receive an MA in English literature at Columbia University and a PhD from the Committee on Social Thought at the University of Chicago, where I studied philosophy, music, and literature.

After completing my course work at Chicago I became a music critic for *The New York Times*, then music critic for *The New Republic*, and in 1991, Chief Music Critic for *The New York Times*. Beginning in the fall of 1995, I began work as cultural critic-at-large for *The New York Times*. I could be considered an eclec-

tic intellectual, who turned in his exam papers in algebraic topology, never to return. But the real astonishment, even for me, about my career, is how deeply the study of mathematics has informed its every turn. I was torn, for example, during my years of serious mathematical study, between music and mathematics, and I often rushed to the piano practice rooms straight from the classroom.

That simultaneous attraction became the subject of a New York Times essay in 1981 and led to my book, *Emblems of the Mind: The Inner Life of Music and Mathematics* (Times Books). I attempted to write for the educated lay person about the essence of each activity and show why they are so closely related.

Even when studying poetry or philosophy, I found myself thinking in a "mathematical" manner. By mathematical I do not mean arithmetical, but full of love of analogy, mapping, abstraction, and clarity that I learned from my studies at Yale and Brandeis.

Whether analyzing how an image was used in a poem, or determining whether a philosophical argument was rigorously convincing, whether thinking about how a musical composition can have emotional power by combining abstract sounds, or trying to figure out how a novel creates an uncanny sense of reality out of images and words, I find myself, again and again, relying on the discipline and the patterns of thought I developed as an aspiring mathematician. That, at any rate, is something I keep aspiring to as I pursue my non-mathematical career.

Christine Rutch

BS Mathematics
Cedar Crest College

Associate Statistician
Air Products and Chemicals, Inc.

I work in the Applied Statistics Department at APCI. I generally work on short-term projects that can be solved and implemented in less than a month. One of the exciting problems I've helped solve is finding the probability of tornadoes hitting areas across the country. Industrial companies are especially interested in this issue because their facilities usually are comprised of large above-ground structures. This project required me to have an understanding of probability theory, a knowledge of a computer package (SAS) to analyze the data, and the ability to communicate the results in non-technical language to management. The final result was a map of the USA on which the probability of tornado occurrence was color coded.

Another area in which I have become involved is Statistical Quality Control (SQC). Courses in quality control, process improvement, and design of experiments have been especially important to me in my job. Currently our company is on an awareness campaign to inform our community about SQC and how it can be used. I am helping by teaching courses to technicians, operators, and engineers. A training session runs anywhere from a three-hour introductory session to a three-day course. Training others is one of my most rewarding responsibilities.

I have attended some outside courses and learned how to build front-end menus in SAS (a statistical package). Recently I've spent considerable time designing applications in SAS that are menu driven and do not require the user to know a lot about statistics in order to generate statistical reports. One application is called PROBIT

which allows the user to input experimental results showing the toxicity of certain chemicals and generate a model that represents the potency of the chemicals.

I'm also involved in some failure-time analysis of safety valves. There are many valves on systems located in our plants, and it is desirable to know when they should be tested, altered, and/or replaced. Should pressure build up or something unusual occur in the system, we often depend upon these valves to release pressure and prevent system failures. This project involves finding the distribution that best describes the failure times of these valves. Having such a distribution allows us to calculate the instantaneous failure rate and cumulative percentage failures over time. The following steps are involved in this process: obtain historical data that show when valves failed under testing; assume a distribution (Weibull, exponential, etc,); produce a hazard plot to determine whether the distribution is adequate; use maximum likelihood to estimate the parameters; plot instantaneous failure rates and percentage failure over time (based on the parameters estimated); and finally present the results to the corporate safety group.

One major obstacle I've run into on the job is communicating results to others. I've never taken a technical writing course, and I am feeling the effects of not having done so. I urge everyone who plans to enter a technical field to take communication courses in order to develop the ability to relay information to non-technical colleagues, management, and customers.

Bonita V. Saunders

BA Mathematics
College of William and Mary
MS Mathematics
University of Virginia
PhD Mathematics
Old Dominion University
Mathematician
National Institute of Standards and Technology

While a student at William and Mary, I originally planned a high-school teaching career, but, after student teaching my last semester, I decided to go on to graduate school instead. After receiving a Masters from the University of Virginia, I taught mathematics and computer science at Norfolk State University and Hampton University for three years. After going on for a PhD in computational and applied mathematics at Old Dominion University, I decided to pursue a career outside of teaching. That led me to my current job as a research mathematician at the National Institute of Standards and Technology.

The National Institute of Standards and Technology (NIST) is a government agency that conducts theoretical and applied research to support its goals of advancing scientific technology, improving measurement standards, and strengthening the competitiveness of US industries in the international market. NIST is composed of several technical laboratories. I am a member of the Computing and Applied Mathematics Laboratory. Members of my division serve dual roles as consultants to other NIST scientists while also conducting research in their own particular areas of expertise.

Currently, I am doing research in a field called grid generation. When an engineer designs an airplane wing, he or she has to solve equations that model the air flow over the wing to determine the best wing shape. Since these complicated equations must be solved at many points around the wing, the calcula-

tions must be done on a computer, and the points must be chosen carefully so that the engineer receives an accurate picture of how fast and smoothly the air flows. The points make up what is called a grid, or mesh. Grid generation is the development of computer codes that automatically pick the mesh points. It is used in any area of research, such as aerodynamics, electromagnetics, and crystallization, where equations must be solved on an oddly shaped region. In my work I use techniques from several areas of mathematics including calculus, differential geometry, and numerical analysis.

One of my job requirements is to present the results of my research to other scientists by publishing papers in scientific journals and to present talks or posters at technical conferences and workshops. Getting research published can be both a challenging and rewarding experience. Attending scientific conferences gives me an opportunity to travel within the United States and, occasionally, abroad. It allows me to talk and collaborate with other scientists who are engaged in similar research.

Since I wrote the profile above for the first edition of 101 Careers there have been many changes in my laboratory at NIST, the biggest of which has been the merger of my lab with the Computer Systems Laboratory to form the Information Technology Laboratory (ITL). A lot of the research in my division remains the same, but many of us are branching out into areas which fall more directly under information technology, including the development of test methods, models, tools, reference software, and computer generated reference data.

I have continued to do some work in the area of grid generation, but I am also working on two new projects. One involves the development of numerical software in Java. The other project is the Digital Library of Mathematical Functions (DLMF), a massive undertaking to update and expand the National bureau of Standards (NBS) Handbook of Mathematical Functions and disseminate it in digital format on the web. First published in 1964 by NBS, the predecessor organization to NIST, the handbook contains definitions, formulas, and tables for all types of mathematical functions ranging from the elementary exponential and trigonometric functions to special functions such as legendre, Bessel, and Airy that arise in the mathematical and physical sciences.

I have taken the lead in the development of dynamic 3D graphics and visualizations for the DLMF. Like the original handbook, the chapters of the DLMF will be written by several world-renowned experts in the field of special functions. Creating appropriate graphics for each function will require me to work closely with the authors, who are located in various parts of the US and Europe, but I must also enlist the aid of specialists at NIST who know a lot about scientific visualization and the implementation of 3D graphics on the web. I have already experienced some of the challenges of this task in developing graphics for a chapter on Airy functions for a mockup version of the DLMF.

My knowledge of grid generation has been very useful in doing the visualization work. Although many commercial packages produce adequate surface plots, rescaling a plot to emphasize interesting features of a function may produce an unsatisfactory clipping of the surface. Some packages simply reset values above a certain height to the same constant, producing the shelf effect seen in the graph of Airy function $|\mathrm{Bi}'(z)|$ shown in Figure 1.

By restricting the computational domain to a grid I constructed using contour information about the function (Figure 2), I was able to create the smoothly

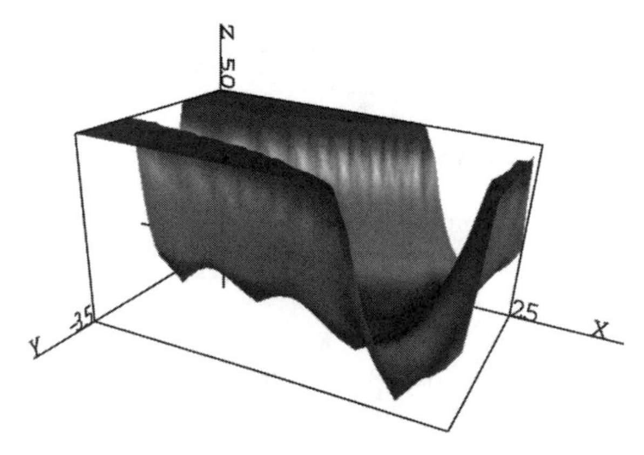

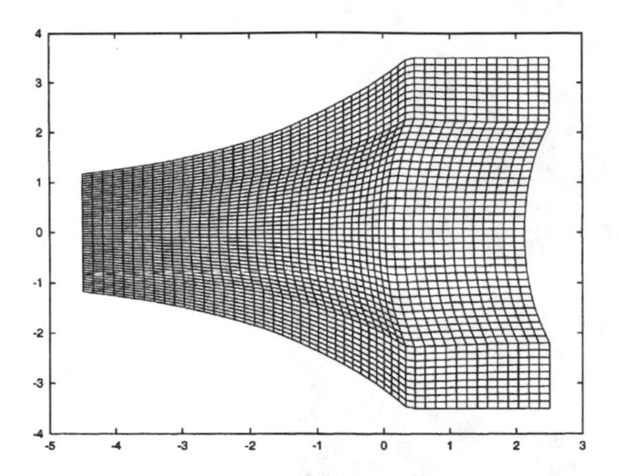

clipped surface shown in Figure 3. Currently I am developing complex multi-connected grids to efficiently clip special functions whose domains contain holes.

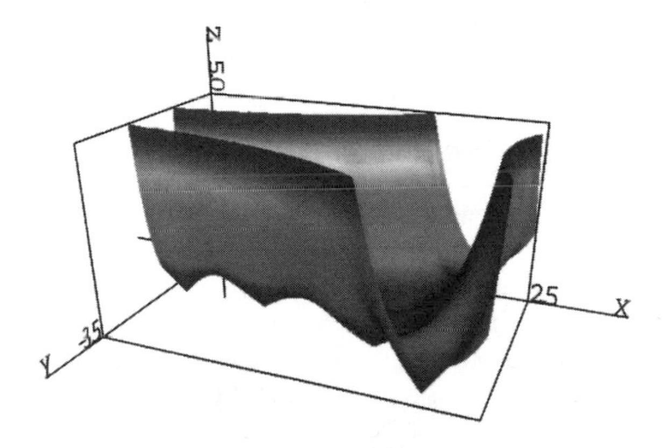

So what lessons have I learned abour working in a place like NIST? First, you must be flexible. Carving out time to keep abreast of your research is important, but you also must be willing to learn new things and work on projects that support the mission of the laboratory. Even though the work may appear to be completely unrelated to your research, sometimes you may find applications where your work can be used. Working on various projects with several people gives you more exposure to the types of work going on and may open the door to other research opportunities. Above all, you must have respect for and be able to communicate well with mathematicians, scientists, and other coworkers.

Do I still love it here? Yes!

Julie Scarpa

BA Mathematics
Hunter College

MA Human Development
Fairleigh Dickenson University

Mathematics Product Manager
Prentice Hall School Division

As long as I can remember I've loved mathematics. As a student I always felt a sense of satisfaction and achievement when I was using mathematics, and I was fortunate to have a few great mathematics teachers who helped nurture my curiosity and love of the subject. Therefore it seemed natural for me to become a mathematics major in college. I almost switched majors when I encountered college calculus, but two things kept me on track — my insatiable interest in mathematics and the fact that all other majors required lots of term papers!

After earning a BA in mathematics I became a high school mathematics teacher, teaching in New York and New Jersey — everything from general math to advanced placement and computer science. During 18 years of teaching, I always took the opportunity to teach new courses and write new curricula, and I tried to convey the power of mathematics to my students. The expanded use of technology in the classroom motivated me to learn to program and use software in order to teach my classes more effectively. This experience proved to be an invaluable asset in my new career.

After years in the school environment, I looked for another way to use my mathematics background. I am currently a Product Manager for Prentice Hall

School Division. My responsibilities include research analyses for new products, development of new mathematics programs, sales forecasting, creating promotional pieces, supporting the sales force (via competitive analyses and sales presentations), and budgeting. As part of the marketing team, I use my knowledge of mathematics and my teaching experience to develop secondary mathematics programs such as our new Algebra 1, Geometry, Algebra 2, and Trigonometry programs.

Knowledge of mathematics is necessary not only in the creative areas of my job, but in the practical areas as well. Making sales forecasts is an annual project, as are planning and tracking the departmental budget. Analyzing, interpreting, and communicating market data are required steps for approval of all new projects.

One of the exciting aspects of my career is that no two weeks are ever the same. I might be traveling with a sales representative to help market our products and formulate new selling strategies, moderating focus groups in order to determine the needs of our customers, or brainstorming with the editorial staff in order to develop a new product. A Product Manager interacts with every department (editorial, production, art & design, advertising, sales support, and manufacturing) in order to perform the job effectively. My career in marketing has given me the opportunity to apply my mathematics skills in a variety of ways.

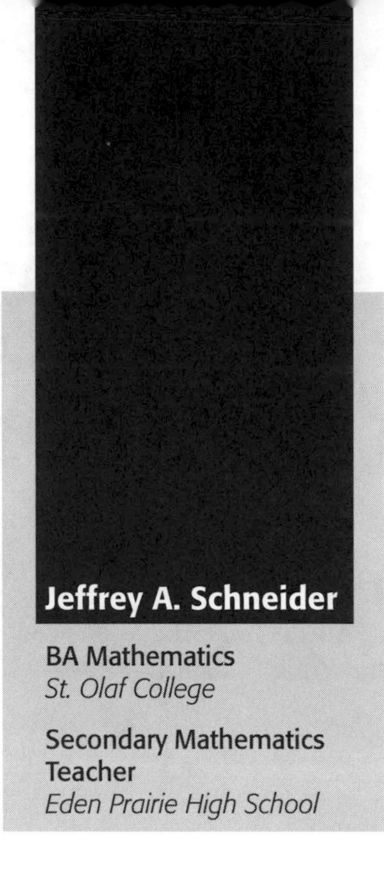

Jeffrey A. Schneider

BA Mathematics
St. Olaf College

Secondary Mathematics Teacher
Eden Prairie High School

I graduated from St. Olaf College in May 1993, with an undergraduate degree in mathematics and education. I applied for a secondary mathematics teach ing position in over 60 different schools in Minnesota and Wisconsin. I had seven interviews with various schools before I was offered a job with Eden Prairie High School in Eden Prairie, Minnesota.

My first year of teaching consisted of three different levels of algebra at both the high school and the middle school. In the following two years, I was able to teach an upper level enriched geometry class along with a lower level applied algebra class. Teaching at both ends of the ability spectrum has been very rewarding and creates a pleasant balance in my day. I enjoy the challenge of both the different students and the different mathematical ideas.

The biggest obstacle I have to overcome in the classroom is trying to present the material in a way that catches the students attention and is relevant to them. The majority of the students in applied algebra either dislike math or have difficulty understanding it. Most of my teaching becomes motivational rather mathematical. Focusing on the student is much more critical than focusing on the pure mathematics. On the other hand, the students in enriched geometry are

much more focused. They yearn for the pure mathematics and its relevance to the higher levels of math.

The best advice I can give to a future teacher is: Make sure teaching is the profession for you. Don't use teaching as a last resort. Teaching math has been a challenging experience for me. There are times when I thought that maybe I should look into another profession that wouldn't be so demanding. After much contemplation, I realized that if you want to do well in any job, it will be demanding. Besides, the rewards from teaching are difficult to find in any other profession.

My main goal in teaching is to make students realize that math is not the frightful experience that their family and friends warned them about. In order to achieve this goal, it requires extra time helping students and creating lessons that are meaningful and are relevant. In the end, there is no greater reward than when a student's face lights up and he/she says, "Thanks, Mr. Schneider. This stuff is easy!"

Joel Schneider

AB Mathematics
Franklin and Marshall College

PhD Mathematics
University of Oregon

M. Ed. Mathematics Education
Washington State University

Content Director
Square One TV: Children's Television Workshop

Vice President for Education and Research
Children's Television Workshop

Television production and mathematics are an unusual combination. While television is exceptionally popular, most of the public think of mathemat ics as something for only a few special people. Square One TV is a daily, half-hour series that is broadcast for children on more than 300 public stations. We use the attractiveness of television to help viewers realize that mathematics is interesting, accessible, and worthwhile. Producers, directors, script and song writers, animators, and many others work on the series. Their job is to make entertaining television pieces. Most have had typical experiences with mathematics. My job is to help them make pieces that are also effective for mathematics.

While I was in school, math came easily for me. I was much busier with music, sports, drama, and books. In junior high school, my favorite reading was science fiction. One novel featured a time machine in the shape of a "three-dimen-

sional snowflake curve." I wondered about that and asked my teacher what it would look like. She suggested that I try to figure it out for myself. That started me on my way to becoming a mathematician. In time, I did figure it out and even wrote about my solution in a paper that was published in a math journal.

In graduate school, I specialized in algebra and number theory. I later taught mathematics at the Pennsylvania State University, including the courses for future elementary school teachers. They drew me into math education. I joined a math education program at Washington State University that included a year in the Seattle Public Schools. There I worked with students at every grade level. My experience in the schools became invaluable when I moved on to a curriculum project in St. Louis where we developed an elementary school mathematics program based on problem solving. I wrote geometry lessons and designed workshops for teachers. Later I moved to New York City to lead one of CTW's teams developing educational microcomputer software.

When CTW began to produce Square One TV, I was ready. Because of the nature of our product and the variety of people on our production team, nearly all of my interests and experiences are useful in my work here. Mathematics has always been a rewarding field for me, both for its intrinsic beauty and interest, and for the variety of ways I've been able to work with it.

Square One turned out to be a long and fulfilling project. The show was on the air from 1987 through 1994. We produced 230 programs for broadcast. Later we produced a special series for teachers called *MathTalk*, comprising 20 videos using segments from the *Square One* library, but set in a new format. We also wrote a guide to help teachers use the videos as part of their math teach-

ing. Setting up this school project and guiding its execution called on my earlier experiences in teacher education and curriculum development.

Another television program that grew out of *Square One* is *Risky Numbers*, a math game show that we produce in several other countries. We based it on some of the short games that were part of *Square One*. *Risky Numbers* has all of the trappings of a typical television game show, with the extra attraction that each game involves the contestants in some area of mathematics, such as probability, estimation, or geometry. *Szalone Liczby*, the Polish version, has been on the air for more than five years, and more shows are in the works. *Yi Er San Tse Wu* (*1 2 3 4 5*), the Chinese version, premiered in Shanghai in 1999. Each show features local children as contestants and their schoolmates as audience.

Using media to help children lean is an old idea, made famous by *Sesame Street*, which is the main project of Children's Television Workshop. For the last few years, I've been managing CTW's Education and Research Division. That takes me away from most of the day-to-day production work, but I am able to use my own experience to guide the people who are responsible for the educational substance of our products, especially *Sesame Street*. I've enjoyed my part in *Square One*, *Risky Numbers*, and the other television shows. I've also been able to be an advisor to other informal education projects, most often mathematics projects with media components. My experience shows that an interest in mathematics can lead in unusual and rewarding directions.

William C. Schwartz

BS Mathematics
The University of Chicago
MA Mathematics
University of Missouri
PhD (honorary)
Engineering Science
University of Central Florida
President & Chairman
Schwartz Electro-Optics, Inc.

How has my mathematics background helped me over the years? First, it gave me an entry into the aerospace business which led to operations research and systems engineering. Also the logical thought processes taught in mathematics and my operations research background helped me in business management both in being able to analytically read a balance sheet and income statement and also in estimating the effects of management decisions on business operations.

I taught mathematics during graduate school and for one year after graduation at the Milwaukee School of Engineering but then decided that teaching was not the career for me. I next went to work for one of the military aircraft manufacturers (North American Aviation-NAA, later to become Rockwell and now Boeing) as a systems analyst and performed ballistics analysis and developed gunsight and fire control systems equations. This then led to the evaluation of the performance of aircraft fire and bombing systems. As I continued working for the company I was given more responsibility in areas involving my mathematical background and next became involved in war gaming and operations research.

In the late 1950s I developed an interest in space vehicles and orbital calculations and was a member of the founding group within NAA involved with the development of space vehicles and which developed the Apollo space system. In the early 1960s I moved to the Martin Company (now Lockheed Martin) where I managed three departments, Operations Research, Systems Engineering and Human Factors for this missile developer. It was during this period that I became involved in laser R&D. Shortly after the invention of the laser in 1960 Martin recognized its potential in missile guidance. I then went on to head the laser missile guidance R&D at Martin during most of the 1960s (Martin developed some of the first lasers used in Vietnam).

In 1968 I left Martin and founded a company, International Laser Systems-ILS, which developed and produced most of the laser target designators and rangefinders, used for laser-guided smart weapons, which flew in Desert Storm, the war in Iraq. In 1983 ILS was sold to Martin Marietta and I retired for one year. In 1984 I founded Schwartz Electro-Optics, again a laser company but this time one involved in both commercial and military markets. The company has now grown to about 150 employees and produces laser systems for traffic control, weapons training, aerospace and industrial applications.

Finally I would urge any mathematician interested in management to get involved in community service. I have been very much a community activist and have gained valuable experience and people skills from this volunteer work and attribute much of my management success to this experience.

Deborah A. Southan

BS Mathematics
University of Miami

Scientific Systems Programmer
Bendix Field Engineering Corporation

My background in mathematics has been invaluable to me in the technical and problem-solving aspects of the work experiences I have had.

My first job upon completion of college at the University of Miami was in marketing. I worked for a small software firm and traveled quite a lot. The analytical skills I developed, as a result of my course of study, enabled me to interface with potential users effectively. I evaluated customer applications to decide how our database could best suit their needs. This information aided sales personnel and helped the marketing department forecast future trends.

Even though I enjoyed the fast-paced world of marketing, I felt the need to expand my work experience to include technical applications. I was then employed by Bendix Field Engineering Corporation working on-site at NASA, Goddard Space Flight Center. The problem-solving approaches I learned in school were essential in my day-to-day responsibilities at the Data Capture Facility.

I work in the Time Division Multiplex Telemetry Analysis section — a section responsible for monitoring the quality and quantity of the data as it is being telemetered down from a spacecraft. Following the successful receipt of the data, analysts check the quality of the data against expected known criteria,

check the amount of data coverage to ensure that all the data are received, process the data into a usable format for the experimenters and, for some of the projects, transmit the data to the user community. I was assigned to analyze the Inter Cometary Explorer, the Dynamics Explorer, and the Cosmic Background Explorer during different phases of my training in telemetry analysis. Each satellite I worked on helped me to gain further knowledge and experience about telemetry, data processing, and trouble shooting problems.

Later, I was assigned to the Upper Atmosphere Research Satellite development and analysis team. Prior to the launch of UARS in September of 1991 we were responsible for supporting the in-house efforts of the Software Group, the Acceptance Testing Group, the Hardware Engineering Group, and the Central Data Handling Facility, as well as all network testing and operations readiness testing. I also prepared the necessary tools for launch support, wrote the documentation on procedures to be followed, and aided in the training of incoming personnel.

Currently, I am in another phase of my career development. I have left the workplace to be a stay-at-home mom. I have gone back to school and completed my certification for secondary education-mathematics. I feel this addition to my education expands my opportunities for the future. I am really enjoying my daughter and am thankful for the security my degree in mathematics gives me for future re-entry into the workplace.

Mitchell Stabbe

BA mathematics and political science
University of Rochester

JD Degree
University of Chicago Law School

Partner
Dow, Lohnes & Albertson, PLLC

I attended the University of Rochester and graduated with a Bachelor of Arts with a double major in mathematics (with honors) and in political science. Even while I was in high school, I intended to attend law school. I had always done well at math: I enjoyed the logic and the problem solving aspects of it. Once you solved an equation, you had an answer and you were done.

Once in college, my advisor suggested that I should have some courses on my transcript that showed I knew "how to write a sentence," so, I took a few political science classes. I found them far less difficult, but, in many ways, more interesting than math courses in that they taught me about the "real" world. As a result, I signed up for more and more and, by my senior year, found that I had enough for a double major.

While in law school and, now, in practicing law, I still find myself thinking and reasoning linearly, as I would in solving a math problem. Step A leads to Step B which leads to Step C and so forth. In law, you start with a basic legal principle or proposition, apply the principle to the facts at hand and reach a conclusion.

$f(x) = y$, so to speak. But, in the practice of law, to say the least, the conclusions are not so clear cut.

In particular, I practice primarily in the field of trademark law where virtually nothing is black or white: everything is a shade of grey. For example, a word may be protectible as a trademark if it is considered sufficiently distinctive to identify the source of the product to consumers. But, the same word may be distinctive for one product (e.g., APPLE for computers), but, generic for another (APPLE for fruit). In addition, one mark may infringe another if there is a likelihood of consumer confusion. That question turns, in large measure, on whether the respective marks are too similar considering numerous other factors, such as the distinctiveness of the respective marks, the similarity of the respective products, the channels of trade through which the products are sold or promoted and the level of sophistication of the consumers. The analysis of whether there is infringement is truly a multi-factor, non-linear inquiry.

Thus, particularly in my area of practice, virtually everything is subjective and reasonable minds can differ. Despite the satisfaction that I have always gained from "solving" a puzzle, I have come to enjoy and embrace the ambiguity involved in the practice of law.

Arthur P. Staddon

BA Mathematics
Denison University

MD
University of Pennsylvania School of Medicine

Clinical Associate Professor
Hematology-Oncology
University of Pennsylvania School of Medicine

When I graduated from Denison University, I proceeded to go to medical school at the University of Pennsylvania. I had not followed a typical course in taking the usual premedical curriculum in college. I majored in mathematics mainly because I enjoyed the thinking that mathematics requires, and I found it fun to do. I was interested in applying my skills in math and science with my interest in people and service. I was concerned that I might have difficulty being admitted to medical school, not being a typical premed. I found that doing well in mathematics opened the doors to the very best medical schools. The rigorous discipline in training in analytical thought processes prepared me extremely well for medical school. I found no other courses that were more rigorous than those that I had encountered in my mathematics studies.

After medical school, I finished a residency in internal medicine, then in hematology and oncology. I was extremely well served by my mathematics background because of my ability to think logically and quantitatively. In medicine, one is faced with a problem which must be thoroughly analyzed before a solution can be found. This process is very similar to the discipline of mathematics. My medical training has given me the opportunity to teach in many areas of the

world. I spent a year in Korea and traveled throughout the far east. I was in Iran four months teaching residents. I taught for a year in Egypt at the Egyptian National Cancer Institute. Recently I have participated in education seminars given in the Dominican Republic and Ireland. It has been an exciting bonus to be able to travel and experience other countries and cultures.

After my training, I joined the staff at the Graduate Hospital in Philadelphia, Pennsylvania, as a hematologist and oncologist. I see patients, I teach medical students, interns, and residents and do clinical research in cancer and in AIDS.

All of my work has been well grounded because of my initial studies in mathematics. The ability to think logically and to analyze problems and situations clearly has been invaluable. Mathematics also teaches one to be precise, which is key in research as well as in clinical medicine. In medicine, it is not as important to be brilliant as it is to be thorough and precise in our thinking. Currently, I am the Co-Director of the Cancer Program at the Graduate Hospital. I feel I have been extremely fortunate to be able to work in an area which is intellectually and academically stimulating and also is of service to mankind.

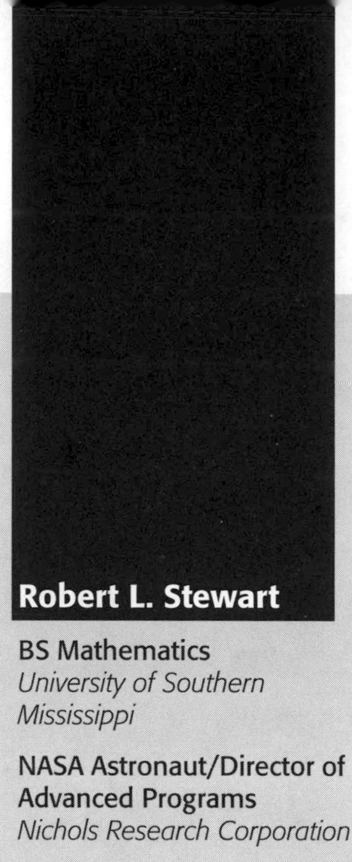

Robert L. Stewart

BS Mathematics
University of Southern Mississippi

NASA Astronaut/Director of Advanced Programs
Nichols Research Corporation

Mathematics can be an end in itself for some people, or it can open a multitude of other doors. The first part of my professional career did not rely heavily on a math background. I flew armed helicopters in Vietnam and was a flight instructor in primary helicopters. But soon the dividends were realized. I was sent to the Army's Guided Missile Systems Officer Course. The heart and soul of the course was applied mathematics, and I was well prepared. Laplace transforms became tools for stability and control analysis, not mathematical abstractions. Calculus of Variations became a means for computing optimum flight profiles. In short, the mathematics took on a concrete reality.

A Masters Degree in aeronautical engineering from the University of Texas at Arlington set the stage for my next step, the US Naval Test Pilot School (USNTPS). USNTPS is an ideal mix of the theoretical and the practical. The fine nuances of the stability equations explored in the classroom in the morning, are graphically demonstrated in the air that afternoon, thus cementing the relationship between the mathematics and the real world.

As an experimental test pilot assigned to Edwards Air Force Base, CA, I was fortunate enough to test five prototype helicopters in four years. The highlight was being the project officer and chief test pilot on the Apache attack helicopter.

The ultimate test of a mathematical analysis is to bet your life on it by flying an aircraft that isn't quite "house broken" yet.

Following a very stiff competition, I was selected as a NASA Astronaut. My background led to assignment as the Astronaut Office representative to develop the space shuttle Entry Flight Control System, a task I pursued for three years. I found that it was harder to bet someone else's life on a mathematical analysis than it was to bet my own.

While at NASA, I flew two space shuttle flights. On STS-41B Bruce McCandless and I conducted the first orbital flight tests of the Manned Maneuvering Unit (MMU), the first untethered extra-vehicular activity from a spacecraft in flight. Being all alone, 1,000,000 feet above the earth, traveling at nearly 17,500 mph, makes one very happy that Isaac Newton and Johannes Kepler were steadfast in their pursuit of mathematics.

In 1987, I was promoted to Brigadier General. Retired from the Army in 1992, I am now Director of Advanced Programs, Nichols Research Corporation, Colorado Springs.

It should be evident that each step in my career has rested on a firm foundation in mathematics. For me, the study of mathematics was the key that opened the doors to the universe.

During my tenure as Risk Control Administrator/Corporate Risk Manager I appeared as the representative of the grocery store chain at Workers' Compensation hearings, at Examinations Before Trial, and even in Supreme Court as part of the defense. I initiated the implementation of the processing and payment of small medical only (first aid) workers' compensation claims through our office instead of through our Workers' Compensation Insurance Carrier saving the company hundreds of thousands of dollars.

From Great American/Victory Markets, Inc., I went to work for Riedman Insurance as a Marketing Specialist and Claims Manager. The Riedman Corporation is one of the 25 largest insurance firms in the nation, and one of the top ten privately held brokerages in the country. Over 100,000 business clients have chosen Riedman Insurance to offer them protection for their financial needs. Through its branch network of over 72 offices in fifteen states, the corporation serves governments, businesses, and individuals alike. I work in the Endicott,

New York office where a team of over thirteen professionals serve clients of all sizes and circumstances.

As a Marketing Specialist I work with the Account Executives to provide a highly analytical, step-by-step methodology that ensures each business the best possible risk management plan. I prepare detailed insurance specifications for the client and market the client's insurance program with the insurance companies we represent. Riedman's national presence results in a unique working relationship with more than 150 different insurers. This ensures that the client receives the most comprehensive and competitive insurance program available. We can also access carriers to insure any international exposure.

As Claims Manager, I help each client establish and implement programs in which the client participates in the payment of workers' compensation medical only losses, more commonly known as first aid claims. I advise regarding accident investigation and reporting, return to work programs, and proactive claims management issues. I also monitor claims activity and provide clients with customized Claims Status and Case Management Reports. I am available for in-person claims reviews, medical bill reviews, and, when needed, will coordinate outside assistance such as referrals to independent medical examinations, and auto glass or body shops. I complete loss notices on behalf of the client and forward them to the respective insurance provider. I then act as a liaison between the client and the insurance company, guaranteeing the client prompt service and the broadest coverage available under their insurance contract.

Kathy Haas Stukus

BA Mathematics
Seton Hill College

Member Technical Staff
AT&T Network Systems

As a member of the technical staff for AT&T, I am responsible for system verification of complex computer operations systems that support specific aspects of AT&T's networks. In this role, I work as part of a product team which includes systems engineers, software developers, and technical support engineers. The team is responsible for the development of a system from its inception until it is ready for delivery to the market. My job involves planning, writing, automating, and executing tests in a laboratory environment, and is the final step in the development of a product before being installed at a customer site. This is a challenging assignment, due to AT&T's commitment to the highest levels of quality and customer satisfaction. It is imperative to discover and correct design deficiencies or defects prior to product implementation, since defects found in the customer environment are extremely expensive to correct, let alone the ill effects they may cause.

The challenge in my job lies in creating test methods and procedures to simulate the actual customer environment in a laboratory testing environment, and uncovering the greatest number of defects in the shortest time possible. This is not a trivial task when one considers the variability of customer use. Doing my job well requires a great deal of creativity, attention to detail, logical thinking, analysis of variables, perseverance, and problem-solving skills. What better train-

ing for this job than a background in mathematics, for these are exactly the skills that those who major in mathematics develop!

My assignments at AT&T have covered all facets of the product life cycle, from software development, customer training, technical support, through customer site management. So you can see that there is a great deal of variety in my job, providing different challenges with each new assignment.

What experiences and training led to my job with AT&T? I look back to when I was a high school student who loved mathematics, and was encouraged by my parents and a dedicated teacher to consider mathematics as a college major. My father believed mathematics to be a good choice for me, since the options for a career were many and varied. How wise my father was! After receiving my Bachelor's degree, I accepted a position as a high school mathematics teacher, taught for a few years, and then resigned to become a full-time mother. After 14 years of staying home to raise my three children, with the cost of their college looming ahead in a few years, I began to think of resuming my career, but wanted to try something different. I enrolled in computer science courses at our local university, finding that my mathematics background made the transition to computer science easy, since the same kind of problem solving and critical thinking skills are required. After completing an internship with AT&T, a job offer followed, and at age 40, I launched a new and satisfying career.

Lisa M. Sullivan

BA Mathematics
Bowdoin College
Senior Budget Analyst
Bureau of Medicine and Surgery
Department of the Navy

I can still see my first grade mathematics teacher as clear as day — flash cards and all. It was then that I fell in love with mathematics. Math was fun! I always thought of it as a game, and even more so when I studied algebra. Throughout my educational career, I enjoyed mathematics and was very fortunate to have excellent teachers who helped me build a solid mathematical foundation. With a very strong high school background in mathematics, I enrolled at Bowdoin College, not sure of what I wanted to study. It was within my first college semester of calculus that I decided to study mathematics. I pursued the algebra and number theory track as a mathematics major and also a major in Spanish.

Upon receiving my BA, I began my career with the Department of the Navy in the Centralized Financial Management Trainee Program. This program is an intense two-year internship run by the Navy Comptroller. As a budget analyst trainee, I was assigned to the National Naval Medical Center in Bethesda, MD. I worked on the initial mathematical model for the Capitation Budgeting Methodology for Military Treatment Facilities. This model included extensive algebraic formulas used to determine the cost of health care on a per capita basis for each military hospital. After successful completion of the trainee program, I worked at the National Naval Medical Center as a budget analyst. Each day mathematics assisted me in performing my duties as a budget analyst, as I worked primarily with spreadsheets in *Lotus*. Even though computers can do most of the calculations needed, the creation of formulas is still the programmer's responsibility.

Currently, I am a senior budget analyst in the Department of the Navy, Bureau of Medicine and Surgery. The Bureau of Medicine and Surgery is headquarters for Navy Medicine, and oversees 4.5 billion dollars in total resources. I analyze the Medical Budget Programs for Navy Medical Treatment Facilities all over the world. I am involved with many unique and mathematically interesting analyses. For instance, I have worked on various cost models outlining the financial future of Navy Medicine. Trend analysis and forecasting are also a major part of my job as there are various performance indicators in place (e.g., ratios such as Patient Care Costs per Medical Work Unit) used to compare like hospitals (e.g., standard deviation from mean of peers). In my job I use algebra, statistics, and also mathematical modeling as in operations research. Moreover, working for the military offers unique mathematical problems which are not present in private industry, such as determining the optimal number of Navy surgeons needed in wartime.

Throughout my educational career I have always been challenged by mathematics, and now I am equally challenged in my career with the Department of the Navy. Mathematics is critical to my job as a budget analyst. Because my job involves mathematical modeling, I plan to pursue a Masters in operations research in order to be able to solve even more complex problems. Finally, mathematics today plays an important role in career paths, as John McLeish wrote: "The history of numbers shows that most advances were made by mathematicians working in the mainstream, relating their work to actual needs (for example, finding ways to predict the flooding of the Nile, to make correct tax assessments or to translate enemy codes in wartime)." *For me, Math is still fun!*

Elizabeth Sweet

BS Mathematics
Loyola College

Mathematical Statistician
Bureau of the Census

In May 1988 I graduated from Loyola College with an undergraduate degree in mathematics. Shortly thereafter I began working at the Bureau of the Census as a mathematical statistician. I do not work on the Decennial Census; rather I work on a few of the ongoing surveys that the Census Bureau conducts.

The survey on which I currently work is the Consumer Expenditures Survey. This survey gathers information from individuals pertaining to items that they purchase, everything from toothpaste to cars. After the data are collected, they are sent to another agency to produce the Consumer Price Index (CPI).

My initial question was, "So how does statistics come into play?" Believe me, it does. The first year opened up a new dictionary of terms: sampling, clustering, variance estimation, and more. It quickly became apparent that there was an entire world of statistics beyond what I had previously experienced.

As a mathematical statistician, my main responsibility is to ensure that the sample selected is representative, so that the "best" data are obtained. This procedure includes projects in sample selection, tracking non-response, testing new questionnaires, and computing variances on the data collected.

We are always researching new ways of collecting accurate data for less money. This sort of cost modeling was the idea behind a very interesting project which

I ultimately presented at an American Statistical Association meeting. I developed, with input from many colleagues, a computer simulation that mimics interviewer travel in a geographic region. Our goal was to decide whether different travel patterns would save travel time, thus saving money. It was a great experience for me to use my computer skills from Loyola days while also gaining experience in public speaking. Although that experience was both exciting and scary, I consider it an honor and look forward to presenting the next paper.

In my spare time, I am working on a Masters in applied statistics at George Washington University. Although it is a lot of work, I find it rewarding to learn something in the classroom and then immediately see the application of that knowledge in my job.

Although the Census Bureau encourages furthering one's education, I am thankful that they do not hold back interesting and challenging assignments from those who do not have advanced degrees. It has been my experience that many of the skills needed for successful work are obtained at the undergraduate level. However, I do look forward to completing my Masters degree and all the new challenges that lie ahead at the Census Bureau.

Julie S. Tarwater

BA Mathematics
Texas Tech University

Quality Control Manager
Biles & Associates

I began my studies at Texas Tech University as a Chemical Engineer. Since this article appears in an MAA publication, you might guess that I did not com plete my studies in that field. My first mathematics class was Calculus I which I took during my second semester of college. My calculus professor, who was also the undergraduate advisor for mathematics, began ruthlessly persuading me to change my major to mathematics. The strong suit of his persuasion was the fact that the field of mathematics offered a career rich in opportunities, especially for women. After a year of indecisiveness, I eventually changed my degree to mathematics. Almost ten years later, I believe that choice was the best career decision I could have made.

I graduated with a BA in mathematics and accepted a job with a mathematical and statistical software company, IMSL Inc., in Houston, TX. The company, now Visual Numerics Inc., provides software to the U.S. Government, many universities and research organizations, as well as various international industries. The IMSL software is embedded in C and Fortran software applications all over the world. I was originally hired based on my mathematics knowledge and problem solving skills despite the fact that I had no programming experience. I served the company as a Technical Support Engineer which included developing and presenting product demonstrations, assisting customers with software questions, and teaching training courses to customers and IMSL employees. This position

allowed me to travel over the United States and to many foreign countries including Japan, Australia, China, England, France, and Mexico.

In 1994, I accepted an offer to work with a process control software company named Biles & Associates. Biles & Associates provides a process data history to many Fortune 100 companies including many pharmaceutical, chemical, and food manufacturers. The software is used to track critical process data in the manufacturing environment. For three years, I taught customer training classes on the process control software. I was able to use my background in mathematics while teaching topics in the area of Statistical Process Control (SPC). My current position with the company is Quality Control Manager. The Quality Control department is responsible for thoroughly testing the product in order to deliver a high quality product that meets the customer's needs.

Due to my background in mathematics, I have continuously found interesting and challenging career opportunities. Whenever possible, I speak on the subject of career opportunities in mathematics to students of all levels. I feel that my decision to seek a degree in mathematics has helped me throughout my entire career.

Debra W. Taufen

BA Mathematics
St. Olaf College

MM Finance and Accounting
Northwestern University

Marketing Manager
IBM Corporation

When I graduated from St. Olaf College I joined IBM as a systems engineer. At IBM, Systems Engineers are responsible for supporting the company's sales efforts. They assist sales professionals by helping to determine which IBM offering best meets customer's needs and by following up after the sale to ensure that customers use the systems in the most effective way. As a systems engineer I worked with some very complex systems and helped customers solve many problems. My math and computer science background was extremely valuable in helping me think through solutions to the problems that customers presented to me.

One of my greatest assets is my ability to think logically and quantitatively. In many endeavors I have used the same thought discipline that I used for solving mathematics problems in school. That discipline is the most versatile tool I have at my disposal. I majored in mathematics because I liked the subject, but I have used my training in more ways than I ever imagined possible.

After working as a systems engineer for a few years, I was asked to assume responsibilities as a marketing representative. In IBM, sales professionals must demonstrate an ability to comprehend and use highly technical information. They have to understand how IBM's systems work, and they must be able to communicate that information to customers. If it is not done well, customers won't buy our products. As a marketing rep, I put my math background to work

again. I was able to work with many different types of professionals, from systems analysts to presidents of corporations. Customers want to know many different facts about IBM's systems, but, before they buy or rent, they want to know whether the system is financially justified. That means showing them that the benefits they gain by using the system are greater than the dollars they will initially spend. This can be a significant challenge when the system sells for millions of dollars. Because it can take years of savings to afford a system, we often use a time value of money calculation to represent future savings in today's dollars. Another area where my math background has helped me is sales forecasting. IBM regularly asks its sales representatives to project their sales for the coming year. This is always difficult, but by applying probabilities to the many sales efforts underway I can produce a reasonable forecast.

Today I work for IBM as a marketing manager, supervising 12 sales professionals. I am responsible for assigning these professionals to their sales territories as well as determining an annual sales objective for each territory. I base my estimates on IBM's offerings, historical performance, economic factors, demand for our products, the skill of each sales professional, and many other factors. I am also asked to forecast sales for my group, manage expenses, evaluate each professional's performance, and justify investments in productivity tools (such as car phones). Every day presents a new challenge, and I constantly rely on my education to help me adapt and respond quickly. I must be able to anticipate and respond to changes as they happen in order to succeed.

Frederick C. Taverner

BS Mathematics
University of California, Davis
MS Applied Mathematics
University of Arizona
Senior Analyst
Daniel H. Wagner Associates
District Systems Engineering Manager
Sun Microsystems, Inc.

In April, 1995, I left Wagner Associates to join Sun Microsystems. Sun, the third largest computer company in the world, was founded in 1982 by four graduate students from Stanford University. The founders had the idea of building desktop computers from off-the-shelf, commercially available components. Every computer would run the UNIX operating system and would have built-in networking capability. In fact, the company name, Sun, is actually an acronym which stands for "Stanford University Network." Sun is now a worldwide $12 billion company with 29,000 employees.

My first job at Sun was as a Systems Engineer in pre-sales for the Computer Systems division. As such, I worked with a set of Sales Representatives and acted as the technical consultant for their sales activities. My duties included understanding all of Sun's different technologies and our customer's computing requirements. I then combine the two to create computing solutions for our customers. My customer base included both very small and very large companies.

A customer wanted to upgrade a set of desktop terminals (called X-terminals) to stand alone desktop computers. This customer had, perhaps, 10 employees and had just received some funding from a venture capital source to expand their operations. I met with the customer, inquired about their desktop require-

ments, and recommended desktop systems that met their needs. The customer purchased about $20,000 in Sun hardware. That customer became Infoseek, the famous Internet search engine company, and now they purchase about $5 million in Sun hardware and services each year.

Another customer wanted to migrate their financial systems from a competitor's systems to Sun. I spent many hours researching and configuring a custom Sun solution to meet their needs. After about a three-month effort, the customer purchased a total Sun solution to run their worldwide business. That customer was Symantec Corporation, the famous maker of personal computer utility software.

In 1999, I received a promotion to District Systems Engineering Manager at Sun. This position requires me to manage and develop a group of Systems Engineers. There are two major parts to this work: the engineers must be trained to provide solid technical solutions to our customers and they must have the opportunity to grow professionally. These objectives not only provide technical challenges, but personal and relationship challenges as well.

My career has taken me many places and allowed me to do many things. I credit both my mathematics and general education program for this success. With mathematics, I acquired the mindset to technically understand the subject I was working with. With my general and broad education, I developed the communications skills which allowed me to understand a given client's needs and then to present solutions in a clear, precise, and cogent way.

Mark P. Terry

BS Mathematics and Physics
American University
Systems Engineer
Trident Systems Incorporated

Upon reaching my senior year as a mathematics and physics major, I was unsure as to what career choice to make. Back then, I would not have believed you if you had told me that I would be a systems engineer today.

As a systems engineer, I work on the design of various systems that are used by the Army or Navy, usually in a combat role. These systems range from communications to automatic target recognition. My work generally consists of defining a way to accomplish a task (such as recognizing a target) and then creating software that does just that. Along the way, there are many obstacles that I must overcome using the skills that I learned as an undergraduate.

The current project that I am working on involves the use of wavelet analysis and neural networks to transmit images using the smallest amount of memory possible. This is important for the military because they want to be able to transmit high quality images quickly and efficiently (such as during reconnaissance missions). Having learned Fourier analysis and wavelet analysis in school, it is interesting to me to use these mathematical techniques to compress and then decompress images. I also enjoy using the matrix reduction techniques from my old text books to shrink matrices that are too big to transmit. I actually use many skills and draw on many experiences that I had as an undergraduate.

During my sophomore and junior summers, I was fortunate to be chosen to do research at different universities. My research involved mostly writing programs

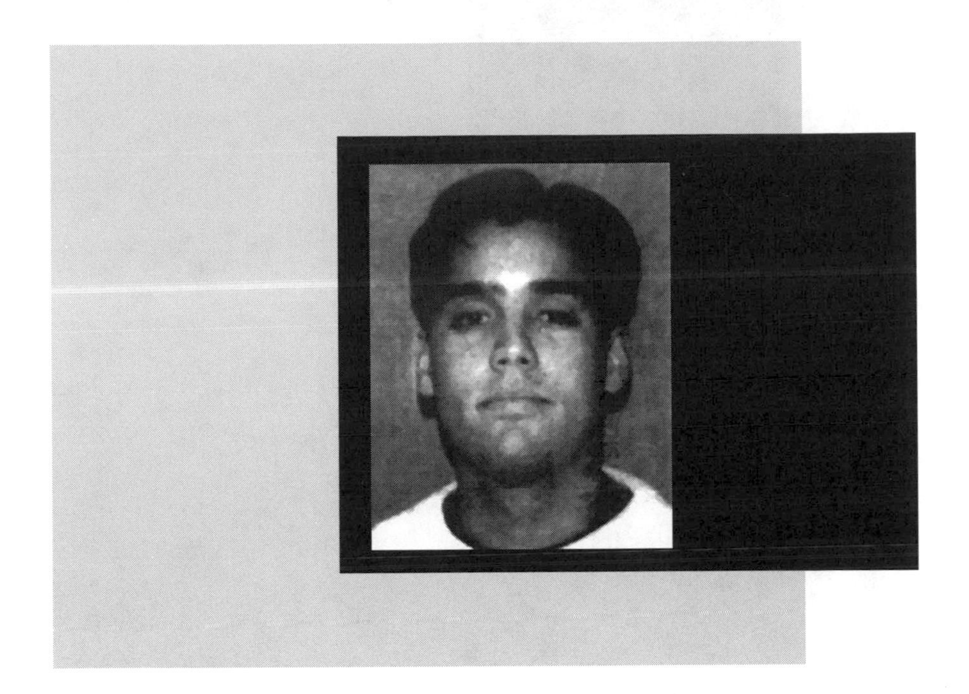

to do analysis of data. These experiences helped me understand how real scientists go about formulating a hypothesis and then check the data from an experiment to see if the theory was correct. I often think back on those days when I try to decide how to check my data for errors.

During my last semester as a senior, I had a light course load, so I decided to get an internship. A graduate student in the department helped me obtain an internship at a systems engineering firm (Trident Systems where I work now). As an intern, I worked on a project that was to produce software that would test the logical correctness of certain inference engines. An inference engine is a tool that makes logical deductions based on given information. The Army uses them to help make decisions such as whether or not to shoot at a target. The tool we produced tested to see if an inference engine followed the rules of the predicate calculus when it processed the information. This was fun because I had just taken a formal logic course the semester before.

When I graduated, the company was happy with my work and they decided to hire me full time. I had had such a great time that I accepted. I have just finished my second year here and I am often surprised to see how frequently my math skills are used.

Jill L. Tolle

BS Mathematics
Loyola College-Baltimore
MS Statistics
Iowa State University

Senior Statistician
I.M.S. America

Early in my senior year at Loyola College I decided that statistics was for me. Being a mathematics major had afforded me the opportunity to take statistics courses as an undergraduate. My advisor convinced me that graduate work in statistics would open many doors and be fun, too. I enrolled in the masters program at Iowa State University.

My graduate level course work enabled me to learn in detail about statistical theory as well as statistical methods, sampling techniques, categorical analysis, and multivariate analysis. My most rewarding experiences came from working as a teaching assistant in the statistics department. I taught basic statistics to business students for two years. I enjoyed meeting and working with my students and made friendships that have lasted past our time in school. I also credit this experience with helping me get over my fear of speaking to large groups. Many times I draw on those experiences when I make presentations or conduct classes at work.

In 1988 I began working for the A.C. Nielsen Company. Everyone is familiar with the "Nielsen Ratings." In addition to that service, Nielsen markets many services to the consumer products industry. Much of the data are collected by scanning

products at check-out counters. I was involved in the analysis of buying trends based on promotions, coupons, and store displays.

In late 1988 I moved back to the East Coast and began working as a statistician at I.M.S. America outside Philadelphia. I.M.S. is a large pharmaceutical market research company which provides many services that help pharmaceutical companies monitor the trends of products in and out of pharmacies. For the last seven months I have been involved with research and development of a projection methodology for estimating the proportion of a drug's volume (per state) that is being paid for by Medicaid. This has become a critical issue for drug manufacturers because of legislation passed by Congress. In addition to this project, I work on an ongoing basis with our promotional testing service to analyze the effectiveness of a company's promotional activity to a select group of test doctors. I am also involved with teaching an in-house basic statistics course to our account/product managers.

My classes and other experiences in graduate school have prepared me well for the problems and opportunities I face every day at work. My work is challenging but also fun. Understanding the necessity of my research as well as seeing my recommendations put into place are a big plus — I don't just sit around and crunch numbers all day.

Lisa M. Tonder

BA Mathematics
Concordia College

MS Statistics
University of Nebraska - Lincoln

Principal Statistician
Medtronic, Inc.

I began working at Medtronic, Inc. in 1991 after finishing my master's degree. Medtronic is a medical-device company with seven different divisions. The division that I work for provides products (pacemakers and leads) that help to speed up a slowly beating heart, a condition called Bradycardia. I have learned quite a bit about physiology working at Medtronic, through courses provided internally and from those with whom I work. An important part of providing statistical support and determining study design is to understand the problems that one faces.

Before products can be sold in the United States, they must receive approval from the Food and Drug Administration (FDA). Careful evaluation of Medtronic device implants through follow-up visits and provocative testing is used to gather data, and statistics are applied to determine product safety and efficacy. My responsibilities are quite varied, starting with the design of a study, designing data forms, writing SAS (computer) code, conducting data analysis, reporting the results, and defending this information to the FDA. I have been part of Medtronic teams that meet with FDA reviewers.

Recently I worked with those who had developed a new generation of pacemaker. The approval of a new type of product is a slightly different process and

more thorough than that required for products that are merely modifications of previously approved designs. This newly designed pacemaker had to be presented and defended to an FDA panel of physicians, and the panel members were permitted to ask any questions that concerned them. My participation required many hours of preparation and rehearsal to try to anticipate the questions and to have answers to them memorized or readily available. The day of the panel meeting finally arrived, and, after a couple of hours of questions, we received unanimous approval from the panel. My job requires me to communicate well, to explain statistical techniques to non-statisticians, and to do quite a bit of SAS programming. The specific classes and skills that have been the most beneficial are calculus, applied statistics, and computer programming.

In December 1996, I was promoted to Principal Statistician. My responsibilities were increased to include supervising the Bradycardia Division's Statistical Group and my secretary. This added a new dimension to my career. These responsibilities include writing performance reviews, helping staff with professional development plans, statistical training, and coordinating the statistical projects. The group is small enough so that I still have time to work on the statistics of clinical studies.

Pacemakers have sensors in them that detect when the heart rate should be increased. The method that we use to test the adequacy of the sensor is to have a patient walk and run on a treadmill with increasing intensity, until they cannot exercise any longer. Each part of the exercise test has a workload associated with it. For the statistical analysis we take the patient's heart rate and normalize it to the maximum heart rate and the workload to the maximum workload. Then, a linear regression (shown in the figure below) is conducted with a "perfect" sensor providing an intercept of 0 and a slope of 1.

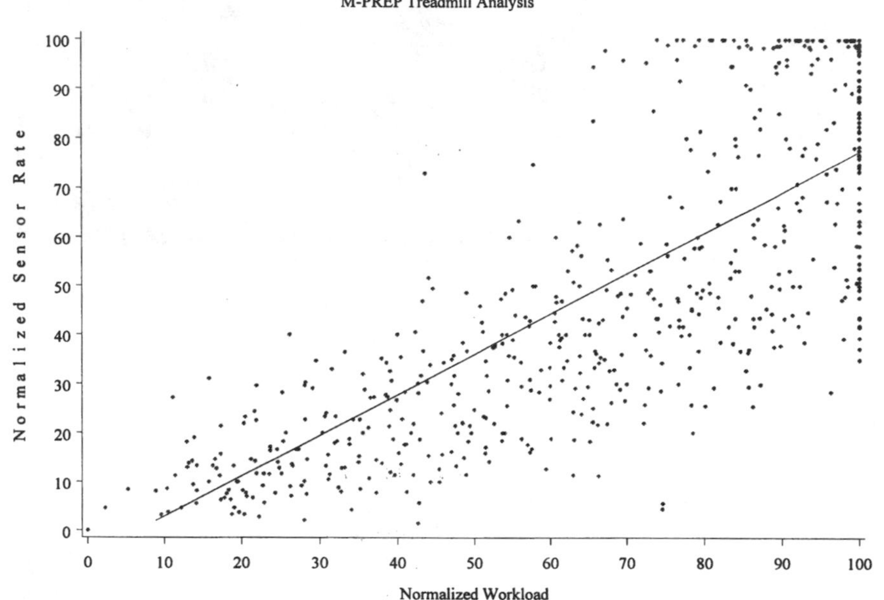

I have also been part of a program to develop the statistical expertise at Medtronic by writing a Clinical Statistics class and teaching portions of the material. This also gives exposure to the statisticians in the company so those who take the class get to know who they should talk to about appropriately designing a study before they start it, instead of giving us the data to analyze after the study has been completed.

In May 1999, my supervisory responsibilities were increased again to include the management of a team that is responsible for the post-marketing surveillance of our leads and pacemakers. There are many things that may be gleaned from the database of information that is currently not being utilized. It is my responsibility to help turn the information into something that is meaningful to others and make sure that people have access to it.

I continue to be responsible for a few projects which require much experience and knowledge of our current products and I am enjoying the added dimension of my supervisory responsibilities. I have recently received a Medtronic Quality award for my work on the FDA approval of our latest pacemaker.

I really enjoyed mathematics in high school and had decided to major in mathematics for my undergraduate degree, with minors in computer science and music. I wanted a career that would combine my interests in computer science and mathematics, but, by the end of my junior year, I decided to pursue a master's degree in statistics. Now, I'm happy that I did so — I highly recommend pursuing a career in statistics.

Sue E. Waldman

BA Mathematics
University of Oregon

MS Mathematics
Oregon State University

Mathematician
Columbia Plateau
Conservation Research Center
US Department of Agriculture

When I left college I had no idea what a mathematician did, so I began my career teaching high-school mathematics, then college mathematics. Later, when home with our small children, I was hired part time by Agricultural Research Service to do "some programming" for a few months. Twenty plus years later, I'm still at the Research Center in Pendleton, Oregon. My job has changed over the years. Originally I wrote programs — statistical analysis, curve fitting or data manipulation and storage — for the entire staff. Today, commercial software is available for many of these tasks which allowed me to return to my first love, mathematics. I am part of a team that describes plant and environmental processes mathematically or "models" them as organized patterns of interrelated theories stated as a series of equations. Eventually I program these equations into computer code to simulate environment and growth. We use such mathematical models to communicate our knowledge of processes and interactions between the physical and biological system of the crop and its environment.

A model provides an answer to a problem or question. Once the question is outlined, we read other scientists' research to become familiar with their published theories; we have "brainstorming" sessions with scientists in various fields of expertise to explore the many aspects of the problem; then many hours are spent making observations and gathering data — weather information, soil infor-

mation, number and sizes of leaves, plant weight, number of seeds on a head, number of roots, etcetera. Our dataset is first analyzed graphically – the dataset is plotted using both the recognized and original theories to decide whether there are relationships with known variables such as temperature, sunshine or rainfall. I transfer these relationships of the observed data to series of equations which is called a model. These mathematical models ultimately provide cause-and-effect descriptions of crop growth and response to the environment. When we build our models using known independent variables, the output from one model can be the input for another. Finally, we bring together all of the processes and transfer them into computer code. Thus, these models become a simulation of a natural process. With a crop growth simulation, a crop can be "grown" in a computer using the current weather information. These simulations can help farmers decide, for example, whether it is economically advantageous to add fertilizer, irrigate the crop, or spray for an insect or disease.

What happens in the field can be observed, but we don't always know why it happened. It is the task of the modeler to determine a cause-and-effect relationship. Many different methods may be tried before we discover what works, but in research we must be willing to make mistakes and recognize errors. I enjoy being a mathematician/modeler/programmer in agricultural research. It requires creativity and problem-solving skills; there are all kinds of things to be discovered and endless challenging questions waiting for solutions. In applying mathematics to the "real world", we observe and measure the world in such detail that we are able to see its simplicity and then communicate its process mathematically.

Rodney B. Wallace

BS Mathematics
University of Georgia
MS Applied Mathematics
Southern University

Systems Analyst
IBM Corporation

NASA's Space Station Program represents a gigantic step toward a new frontier for Man-Space exploration in the 21st century. Not since the Space Shuttle program has there been a major undertaking by NASA. I am a systems analyst at IBM, working on this leading edge of science and technology. Until recently I was a member of the technical staff at AT&T Bell Laboratories, where I spent over three years. Prior to that I was a systems engineer/applications programmer for Singer-Link Flight Simulation. How did my mathematical training influence and enhance my professional career?

After completing my BS in mathematics, I went to work for Singer-Link, a NASA contractor that provided training to the Space Shuttle astronauts. I put my mathematical modeling skills to work immediately, simulating various payloads that were to be deployed from the Space Shuttle. This assignment required a broad understanding of many applied engineering concepts such as Boolean logic, equations of motion, and electrical circuits — all of which have mathematics as a common denominator. Courses such as calculus, differential equations, advanced engineering mathematics, and modern algebra made my assignment a piece of cake. It was very rewarding to see the astronauts out in space working on projects for which you helped prepare them — especially when they thanked the simulators on live television.

After Singer-Link, I decided to complete my MS degree in applied mathematics. With an advanced degree in hand, I decided to apply to one of the most re-

spected research and development facilities in the world, AT&T Bell Laboratories. I accepted a position in the department of performance analysis of computer systems and became interested in queueing theory, which is the analytical tool most widely used to model the performance of computer systems. A queueing theorist studies waiting lines using various theories and tools found in mathematics. Queueing theory provides data on computer performance such as throughput, response time or delay, and utilization. Thus, I was introduced to the world of performance modeling. My fascination with this subject led me to take advanced graduate courses in queueing theory. As a member of the technical staff at the Labs, my assignment was twofold: I developed queueing models to predict the performance and capacity of new products; and I educated our customers on how to properly monitor their systems.

While at Bell Labs I completed one year of doctoral studies in applied mathematics. I then elected to join the Space Program once again by accepting a systems analyst position with the IBM Federal Sector Division. Presently I am leading the technical efforts in computer systems and network performance modeling and sizing for the Space Station Training Facility. This facility will consist of mainframe and mini computers, work stations, local area networks, and actual flight equivalent components necessary to train NASA 's astronauts for their Space Station missions.

My mathematical training has definitely been important to my professional career. With my plan to complete doctoral studies, mathematics will play a very important role in my future.

Martha Weeks

BA Mathematics with a Minor in Computer Science
University of Tennessee

MBA
University of California, Berkeley

Computer Marketing Consultant
Self-employed

Study math and see the world! At least, that's what worked for me. When I was in high school, I was bitten by the travel bug. I've been lucky enough to use my career in the computer industry as a way to achieve my dreams to travel, and a degree in mathematics is the way it all started.

While in high school, I developed an interest in mathematics as well as in travel. I continued to enjoy math courses in college, and I began to realize that jobs in mathematics and the hard sciences are more plentiful and more lucrative than those in the social sciences. In many math classes, computers are used as a problem solving tool, and this got me interested in computer science. I thought that some knowledge of computers could help my after-graduation job prospects, and it did.

Hewlett-Packard (HP) recruited me from college for a position marketing computers. I was responsible for supporting and training sales people. After my first three years at HP, I went back to school for an MBA. After graduating from business school, I was re-hired by HP in Germany and then transferred to Lyon,

France to a start-up division. During my three years living overseas, I traveled constantly on business and was able to enjoy the sights of Europe. I returned to the United States with HP, but, after several years, the travel bug got the best of me again. I took a year off to travel around the world. When I returned, I accepted a position with a small, privately-held company where I sold network management software for two years.

For the past three years I have been working independently as a marketing consultant in the computer industry. Most of what I do as a consultant are things I did when I worked in full-time marketing positions: introducing new products, training sales people, developing seminar programs for customers, pricing new products, forecasting sales, writing white papers to explain new marketing programs or technology, and creating business relationships with strategic partners. Being a consultant is riskier than working in a full-time position, but I have much more control of my own time. I also enjoy the variety of work.

I could have pursued the same career with a degree in computer science, but I am always happy that I chose mathematics as my major course of study. Since computer technology is constantly and rapidly changing, you always needs to learn new things to keep current. These things you can learn on the job. Studying mathematics gives you the tools to analyze problems and think logically, which helps in whatever profession you choose. People have great respect for a degree in mathematics. And, math can help you see the world!

Benjamin E. Weiss

BS Applied Mathematics
Harvey Mudd College

Imaging Scientist
MetaCreations Corp.

I've always been interested in math and science, and in my childhood I was no less fascinated by video games and computer graphics. So it seemed quite natural that I would develop a talent for writing my own computer software, as a sideline of my progression through the world of mathematics. After all, any computer program boils down in the end to applied math and logic, even the ones that draw the prettiest pictures, or perform the most fluid calligraphic brushstroke-based artwork. I began to experiment with simple programs, drawing simple pictures on the screen of my Apple IIc computer.

As I gradually explored new areas of science, more and more artistic possibilities came within my computational grasp. Linear algebra provided tools to flip and rotate graphics in three-dimensional perspective, the basis for countless computer games. Physics and optics opened the door to use the computer as a virtual camera, taking color snapshots of imaginary landscapes and panoramas. All this, of course, as a mere hobby, while I dutifully made my academic way through high school and college.

Then, just after my sophomore year at HMC, my life turned upside-down. In the mundane process of looking for a temporary summer job, I stumbled onto an amazing opportunity; a small startup software company, which was willing to hire me to create computer-graphics programs for them! Suddenly, the roles were reversed. My life began to center around my erstwhile hobby, while my academic career slowly devolved into the sideline. I kept up enough momen-

turn to eventually graduate college with honors, but my heart was in the late nights spent pounding out code, writing new algorithms, and watching all the little pixels fly by.

Since then the company has grown 20-fold, evolving into one of the major players in the industry. My projects include a photo-editing and retouching application called "Kai's Photo Soap" (after Kai Krause, my mentor, and co-founder of the company), and a playful digital-silly-putty program called "Kai's Power Goo," which stretches and twists images to form animated funny-faces and caricatures. I've also done extensive work in more traditional image-processing, plus fractal geometry, 3D raytracing, color theory and optics. Much of this exploration has resulted in collections of colorful toys to play with, but here and there I've also found some serious and meaningful applications for my work.

After graduation, I moved with the company to Santa Barbara on California's central coast, where I live at the beach and go surfing as often as possible. My job gives me complete freedom to develop and follow my ideas, and the nature of my profession allows me to work at home much of the time, and choose my own hours. A good deal of time is spent relaxing, playing piano and looking out the window. The ocean is a source of endless inspiration for me; in fact, my next project may involve a photorealistically rendered, physically-based simulation of breaking ocean waves.

Faedra Lazar Weiss

BA Mathematics-Physics
Brown University

MAHL (Master of Arts in Hebrew Letters) and rabbinic ordination
Hebrew Union College-Jewish Institute of Religion

Research Associate
Girls Incorporated National Research Center

When I read the advertisement for the position of research assistant (the position which grew into my present position), I had three questions. First, though this position looked as though it could have been designed for me, I wondered, would I really have the opportunity to do statistical work, use my research, writing, and editing skills, and support girls and young women in valuing and developing all their interests all at once? Second, I asked, could I juggle full-time work with a husband, two daughters, and a third on the way? Third, I wondered, whoever gets a job from a newspaper ad?

A month later I was editing a report on opportunities and encouragement given to girls to explore math, science, and technology (released on the day my youngest daughter was born). And I was compiling and analyzing survey data on Girls Incorporated affiliates and on the girls and young women they serve. And I was immersing myself in statistics on adolescent pregnancy and substance abuse, data I would possibly use for future publications. Five years later, I still enjoy working on a minimum of three projects at once. Sometimes it is overwhelming, but never boring.

Girls Incorporated (formerly Girls Clubs of America) each year serves 350,000 young people ages six through 18 through a network of local affiliates and outreach programs. At more than 900 sites throughout the United States, girls and young women (and some boys and young men) participate in programs that build their capacity for confident and responsible adulthood, economic independence, and personal fulfillment. As Research Associate at the National Resource Center, one of our two major national offices, I work as part of a team. When we take on a proposal or research project, we share responsibilities: finding information, outlining, drafting, reviewing, and editing. Depending on the project, I may do any or all of these. In addition, I do most of our statistical research, including consulting on survey design, creating databases, and using statistical software to analyze data that we or other organizations have collected. This also requires good writing skills, to explain to all our audiences what the statistics do and do not imply.

My favorite national Girls Incorporated program is Operation SMART™, an initiative to encourage girls to explore and persist in science, mathematics, and relevant technology. I invent and try out interesting and easy ways to show that, given a chance: girls can enjoy math, science, and technology; they can find these areas interesting; and they can realize these areas are critical to most careers. SMART participants learn that, whether you intend to go to college or not, the more math you take, the more career options you have. I know it worked for me!

Michael D. Weiss

BA Mathematics
Brandeis University
PhD Mathematics
Brown University
MA Economics
University of Maryland

Agricultural Economist
US Department of Agriculture

In 1998, after 21 years with the Economic Research Service of the U.S. Department of Agriculture, I left government service to begin the next, and current, chapter of my mathematical career.

I look back with amazement at the fact that, although I was employed by an agency devoted to agricultural economics rather than mathematics, not once was I assigned a subject area in which I couldn't find interesting and enjoyable mathematical content. Federal government agencies seem to provide researchers more freedom than the private sector to choose projects or to steer assignments in a compatible direction. In this sense, federal government employment perhaps occupies a middle ground between academic and private-sector employment.

My experience over the years has convinced me that mathematics can be found wherever you look. For example, when I was working in a research group on food-safety economics, I noticed that a probability theoretic argument could be used to show that larger fast-food chains stand to lose not merely *more* revenue from foodborne disease incidents, but a higher *percentage* of their revenue. This finding was reported in the food industry press. In another study, I drew on

generalized central limit theorems to evaluate safety data on heat-treated meat patties. My conclusions—in disguised form—actually found their way into the Congressional Record. There's a lot of mathematics in hamburgers!

My final government project concerned precision farming, an emerging technology that allows farmers to vary the application rate of crop inputs like fertilizer according to how much is needed at each specific location in the field. Using the symbolic mathematics software *Mathematica*, I developed a computer model that represented the level of in-ground or applied fertilizer as a surface lying over the field. Agronomic rules for fertilizer application were interpreted as operators on a space of surfaces. Aggregates such as total fertilizer applied were viewed as functionals mapping surfaces to numbers. When the farm field was treated as random—an approach that was actually simulated within the computer—aggregates even took on the role of random variables. The model allowed me to draw empirical conclusions about the economic and environmental viability of precision farming.

My experience using *Mathematica* was so rewarding that, after leaving the government, I became an independent *Mathematica* trainer and consultant. This activity has provided a perfect outlet for my love of mathematics, teaching, computers, and *Mathematica* itself. It brings me into contact with highly trained and motivated scientists and engineers, all the while drawing on my knowledge of both *Mathematica* and mathematics and providing me with an ever-growing list of subjects in which mathematics plays an important role.

Gary L. Welz

BA Philosophy and Mathematics
M.Sc. Mathematics
Bedford College, London University

President and Founder
Science Television Company

Director of Corporate Business Development
THINKNewIdeas

Mathematics and all the sciences have entered a remarkable new phase in their brief history. Until about 1450, scientific information was writ ten by hand one copy at a time. Then the printing press put hundreds of copies of the latest scientific books into circulation with drawings of plants and other specimens recorded faithfully in all copies. Since the 1970s, scientists have begun to communicate through the circulation of films, videotapes, and computer animations. Soon, television will be used by scientists to communicate their discoveries to the entire world instantly and with the enormous visual advantage of animation – moving pictures! It is my great pleasure to participate in this historic change.

My first love in mathematics was logic. I came to the subject from philosophy because I loved the certainty of mathematics and its rigor. When I graduated from Bedford College in 1977, I decided that I'd had enough certainty and entered the very uncertain world of show business. I spent the next 12 years devoted to acting, directing, writing, and producing for both the stage and screen. In the mid-80s I was astonished by the beauty and significance of the computer graphics that were being created by mathematicians in their research. That gave me the inspiration, or should I say obsession, to help ferry mathematics and

other sciences into the television age. Ever since, I have enjoyed bringing together the two very diverse worlds that fascinate me.

In 1989 I founded the Science Television Company to produce videotapes with mathematicians. Among the titles are *Chaos, Fractals, and Dynamics* with Robert Devaney, *The Beauty and Complexity of the Mandlebrot Set* with John H. Hubbard, and *Natural Minimal Surfaces via Theory and Computation* with David Hoffman.

In 1990 I proposed the creation of a television network for scientists. With the help of a generous grant from the Alfred P. Sloan Foundation this led to the founding of the Science and Engineering Television Network (SETN) . SETN is a consortium of scientific societies that use television and the multimedia publishing capabilities of the internet to communicate richly visual scientific information to scientists, teachers, students, and laymen all around the planet.

During the past five years I have worked in various internet industries. Beginning as a journalist for such publications as Internet World and WebDeveloper.com, I wrote about multimedia technology and publishing on the internet. Later I created web sites and consulted to such clients as IBM, the New York Public Library, Viacom, and the Association for Computing Machinery. I was a Business Development Manager for Rare Medium, a major intereactive services company. My latest job, with THINKNewIdeas (another interactive services firm), involves selling design, technology, and web strategy services.

Dan C. White

BS Mathematics
University of Santa Clara
MS Mathematics
University of Illinois

Actuary

How would you like to work in an area that most people know nothing about? That is one challenge that awaits an actuary. An actuary is a risk mathematician, one who solves real-world problems involving money (sometimes billions of dollars), probabilities, and future events. Most actuaries work for insurance companies or consulting firms; relatively few work for the government or in education.

Here are some questions that actuaries answer:

- How much do we pay for insurance — for life, auto, home, or health?
- How much should we pay for Social Security and Medicare benefits? Is Social Security going broke? Should Social Security taxes be reduced?
- How much should employers contribute to pension plans so that there will be sufficient funds to pay benefits that have been promised?

I am a life actuary by training and in my work. Other actuaries pursue a career in the casualty actuarial field, working with such issues as earthquake insurance and liability coverage for doctors.

At the University of Santa Clara, I explored the broad career opportunities available to mathematics majors, including teaching, computing, and the actuarial field. I supplemented my mathematics program with courses in the liberal arts, economics, and business. Then, at the University of Illinois, I discovered that my

interests were more in solving practical problems than in proving theorems, so I looked into the possibility of a career in actuarial work.

After graduating from the University of Illinois, I worked for an insurance company for three years and took actuarial exams that led toward my Associateship. To become an actuary, one must pass a series of examinations that include college mathematics (calculus, probability, and statistics) and then on topics of special importance to the actuarial field (accounting, taxation, and law). One of my early projects with the company was to establish the amount that people should pay for life insurance policies, measuring profitability (or loss!) of those policies and determining how much money we should hold to pay future death claims. The company's performance depended on how well I did my job.

In my studies I had learned about the field of health benefits and wanted to give that field a try. I joined an employee benefits consulting firm, where I have worked for 16 years. Here I have helped determine pension contributions by Fortune 500 companies, designed special benefit packages for famous executives, and advised a western state whether it should seek external insurance for its medical plan, a plan that covered over half a million active and retired employees. The possibility of working on such large scale projects continues to make my career an exciting one.

Richard Mahaffey White

BS Applied Science
Miami University

MS Statistics
Miami University

Director, Operations Analysis & Research
The OAR Corporation

Mathematical statistics is the foundation of my consulting practice founded in 1978. During my active duty naval career as a cryptologic officer, in addition to shipboard duty, I worked as a systems configuration architect in Command, Control, Communications (C3) and Intelligence. Command and Control involves the collection, correlation, and dissemination of geographical data on friendly, potentially hostile, and unknown air, surface, and subsurface contacts. Positional data comes from cooperative ships and aircraft sharing tactical information from real time fleet sensors and intelligence sources.

Time sequential positions from ships and aircraft are organized into tracks. The initial probabilistic problem is to associate the incoming position update with the most likely track in a near real time processor; the goal is to maintain a variety of tactical displays in command centers both ashore and within afloat and airborne units of the fleet. The second problem is to maintain data base commonality among the numerous participants, linked via satellite communications. Operational fleet commanders exercise disposition of forces and targeting authority based upon near real time information displayed geographically. The third problem is to evaluate tracks and to estimate future positions with statistical confidence. C3 data is accessed via animated computer displays at current or future times, which are stochastic functions of constantly changing track histories.

Enlisting in 1965 as an electronics technician, I received valuable training in radar and the Navy Tactical Data System used to track and control ships and aircraft. In my 27-year career, I have deployed in numerous aircraft, cruisers, aircraft carriers, and battleships. In addition to stochastic modeling, I train senior staffs, conduct afloat surveys, and help determine fleet operational requirements. I received an appointment as C3 Advisor to Commander Sixth Fleet in the Mediterranean Flagship (89-90 including the Bush-Gorbachev Sea Summit).

In assignments at Booz Allen & Hamilton, Inc., a management technology firm serving hundreds of commercial and government clients, I served on the Admiral's staff embarked in the aircraft carrier USS America (Indian Ocean 80-81) and was Battleship Battlegroup Advisor to the Captain, USS Iowa (North Atlantic 86-87). Mathematical modeling plays a major role in Booz Allen's federal and industrial support. I am privileged to have worked for Lockheed Missile & Space Corp., McDonnell Douglas Astronautics Corp., and NRL.

While Officer-in-Charge of the Cryptologic Detachment on the USS Josephus Daniels, I managed afloat data correlation and Over-the-Horizon flight path simulation for Tomahawk Cruise Missiles. At a Navy lab, I used Markov simulation for a signal search received with finite machine states and estimated transition probabilities; a Bernstein Uhlenbeck model for Aerostat station keeping in the upper atmosphere; geographic models for positional data; queuing models for data link communications applications; and astrodynamic models to stimulate satellite RF doppler effects for Time Difference of Arrival calculations.

Paul Willoughby

BA Mathematics
St Mary's College of Maryland
MS Computer Science
Florida Institute of Technology
Lead Software Analyst
PRB-COMPTEK Systems

When I was a child, I was intrigued by the idea of working with computers but I never dreamed I would end up with a degree in mathematics. Throughout my primary education, I found math to be tedious and boring and declined the chance to take calculus in high school. I ended up going to St Mary's College of Maryland. At the time, St Mary's did not have a computer science major; computer science was part of the math department and you had to be a math major to study computers. At first, I didn't like taking all of the math classes. However, the more classes I took, the more I began to actually like the subject. Choosing to be a math major is something I will never regret.

How does a solid mathematics background help those interested in becoming software or computer engineers? First, all computer science theory is grounded in mathematics, particularly discrete mathematics. Second, writing software requires a complete understanding of logic. This is also found in mathematics. Third, and most importantly, most real world problems that we try to solve using computers are described in mathematical terms. Generally, developing software to solve real-world problems involves the following process: analyze the problem space and define the requirements of the solution, translate the problem into mathematical processes, code it, verify and validate it, and release it to the user. For example, a user may want a computer model of the way a particular missile flies in 6 degrees of freedom. To solve this problem, you must first

analyze some actual flight data of the missile, devise a mathematical model of the test data — in this case probably polynomial equations that describe the flight of the missile. Then you must translate it into computer code with a user interface, test it and finally release it to the user. Although, this is a simplified description, it shows one example of how mathematics is an asset to someone in the software engineering field. If you were a programmer with very little or no mathematics background, you would not play a big role in the solution of the problem. Someone else would come up with the solution, hand it to you and say, "here, code this." To me, it's much more challenging and fun to be a part of the whole engineering process.

Analytical skills are not the only attribute necessary to have a rich and rewarding career. You must also have the ability to work with others as a team. Take every opportunity in college to work as a part of a group and develop your leadership skills. It is also necessary to be able to communicate with your colleagues and your customers. To this end, develop your written and verbal communication skills.

Sandra L. Winkler

BS Mathematics
University of Michigan-Dearborn

MSE, Industrial and Systems Engineering
University of Michigan-Dearborn

Research Scientist
Ford Motor Company

My first job out of grad school was in the Environmental Modeling Group of the Chemistry Department, Ford Research Laboratory, developing and applying air quality models. Air quality models are huge computer programs that simulate the transport of pollutants in the lower atmosphere and the chemical reactions the pollutants undergo. I apply the models to gain insight into the effect of increasing or decreasing emissions from power plants, factories and, of course, cars and trucks. Because the models are very complex, I also do research to develop more accurate and efficient simulation techniques and algorithms.

The model development process is a great way to apply mathematics. Every day I apply techniques I learned in many classes. I've used methods from numerical analysis to accurately solve the ordinary differential equations that describe the chemical reactions, and to model the transport of the pollutants, I've included methods from applied linear algebra.

I've also put my minors to work. I use my computer science background daily as I write code for the algorithms I'm developing and testing. Knowing how to program efficiently is very important because these huge models are very

computationally intensive. I combine the math and computer science when I write code that implements the mathematical techniques used to solve the equations in the model. I have applied my statistics minor, too, in the analysis of the output from the models and in the evaluation of air quality standards for pollutants regulated by the government.

I obtained an MSE in Industrial and Systems Engineering after becoming interested in the field while taking an operations research class as a cognate course during my senior year. The systems topics have proved particularly helpful in projects I've worked on involving applications of risk analysis (health effects of ozone and particulate matter air quality), optimization, simulation, and forecasting. Air quality modeling allowed me to apply all my skills and help clean up the environment!

In what seems like a drastic career change, I moved from air quality modeling to business modeling reseach. But the connection is modeling. Now I develop mathematical models based in economic theory rather than chemical and physical processes. While I continue to write code and do statistical analysis on data and results, I do so using methods from risk analysis, optimization, simulation, operations research, and forecasting that I learned while obtaining my masters.

One business modeling project I worked on involved both optimization and risk analysis. The project's goal was to optimally trade off between the risk and return associated with ever-changing foreign currency exchange rates. Another project involved modeling investment in research and development as stock options.

Ana Witt

BS Mathematics
Missouri Southern University

MS Applied Mathematics,
Southern Methodist University

PhD Applied Mathematics
Southern Methodist University

Assistant Professor
Austin Peay State University

Being a professor of mathematics, I obviously use mathematics daily. What may be surprising is the fact that I use mathematics not only in the classroom but also in my personal research. My field of specialization is numerical analysis, especially those aspects dealing with the numerical solution of initial value problems in ordinary differential equations. Working at a university has allowed me the opportunity and freedom to work on research topics which I personally find interesting. A position in industry would not have this flexibility; an employee in industry must work on the projects which are beneficial to the employer. While many students are aware of the teaching responsibilities of their professors, they might be unaware that the research aspect of my career is every bit as rewarding. I have researched and published papers concerning the selection of an appropriate step size when numerically solving initial value problems. Publishing papers and giving talks at mathematics conventions on results I have obtained are a few of the perks of doing research.

Ever since high school I knew that I wanted a degree in a science related field. I had always been good in math and enjoyed related subjects. When I began college, my declared major was engineering. As my understanding of math-

ematics grew, the engineering courses became much easier. This prompted me to seek a greater understanding of mathematics, to the point of changing my major. As an undergraduate, I took a variety of mathematics courses, most of which were in applied fields: statistics, operations research, modeling, and numerical analysis. This background is very useful now, as it enables me to teach a variety of courses.

Teaching at the university level is very rewarding. I enjoy working with students on classroom material and advising them on career options. Many of my students are not mathematics majors, but require mathematics for their disciplines: engineering, physics, premed, economics, biology, and chemistry. Because of my applied background, I am able to generate real-world examples that relate to their majors. One of my students who is a chemistry major developed a model for estimating the optimal amount of surfactant (soap) needed to penetrate the surface of water. A premed student researched a model that describes the manner in which a drug enters and exits the bloodstream of the human body. It is rewarding when students recognize the power and utility of mathematics rather than thinking that mathematics is only a tedious manipulation of symbols.

Much pedagogical research is being done to study the effects of new techniques and technologies. Schools are beginning to require graphing calculators, and many courses have a lab component that requires the use of computers and calculators. I find it exciting to be on the forefront of such changes.

Larry Wos

BA and MA Mathematics
University of Chicago

PhD Mathematics
University of Illinois

Computer Scientist
Argonne National Laboratory

In the early 1950s, I was a grad student at the University of Chicago, considered one of the best in the world, and, to complicate things, I was blind. So we understand clearly, by blind I mean that I could see sunlight. Moreover, none of the material for studying abstract mathematics was in Braille, and no society existed for recording material. I did find a group who put some things into Braille, although we had to invent the notation because Nemeth code, for example, had not yet been contemplated.

After receiving my masters degree, I went on to receive my PhD at the University of Illinois and, in the process, began my study of computer programming. This study led me to apply to the Argonne National Laboratory in 1957, and I am still there. I was given the job of consultant — to figure out how mathematics could enable scientists to get assistance from a computer. What a marvelous opportunity — and it eventually led to my current career.

I invent strategies that enable a computer to solve problems that require deep reasoning. Yes, there do exist computer programs that reason, and reason logically. Such programs prove theorems from mathematics, design circuits, and prove that some computer programs do correctly what they are intended

to do — but not nearly so well as scientists would like. My job is to find ways to make automated reasoning programs more powerful.

At this point you might be skeptical that a computer program can reason. If so, you might read "Automated Reasoning: Introduction and Applications" or "Automated Reasoning: 33 Basic Research Problems." I wrote the first book with my colleagues and the second book alone. I get great pleasure from being an author, especially since readers say they enjoy the books and learn from them.

Now if you were hoping for a stronger motivation for why mathematics and computer science is so exciting, here it is. Much of problem solving that succeeds requires careful, meticulous, logical reasoning. If we can design a general-purpose computer program that reasons as effectively as the better mathematicians, then the importance will be immeasurable. And that is my goal: to develop a program that draws 10,000 conclusions a second, that runs on parallel processing computers, that self-analytically determines how to change its attack on a problem, and that can assist in finding and correcting mistakes.

Blindness is a handicap, just as having a poor memory is. Certainly if I could see the pins, I would bowl far better, but a 140 average isn't too bad. I have a great wife, many friends, and a wildly exciting career.

Sarah Adams Yeary

BA Mathematics
Spelman College
Staff Systems Analyst
Lockheed Martin
Technical Lead
Lockheed Martin Air Traffic Management

Since 1983, I have worked on worldwide air traffic control projects. My first assignment was as a schedule analyst, where I applied probability, statistics, and logistics to determine whether a project was on schedule. I soon moved into a more technical area, system test and evaluation, in which we simulated the flight of aircraft under a wide spectrum of environments to test whether the computer system was functioning properly. I was responsible for looking at the system from an air controller's point of view.

Currently, we are involved a major upgrade to the air traffic control system, the Advanced Automation System (AAS). One significant component of the project is to replace the displays where aircraft are viewed by the controllers. The new Display System Replacement (DSR) will display more information to the controllers and decrease the likelihood for human errors. The first step was an interim replacement system, Display Channel Complex Rehost (DCCR). The AAS is expected to carry air traffic control operations into the 21st century.

The Display Channel Complex Rehost (DCCR) project was successfully completed ahead of schedule and under budget in early 1997. The Display System Replacement (DSR) project was deployed to the Chicago Air Route Traffic Control Center (ARTCC) in 1998 and is currently in operational use. I was involved

in the software testing using simulated aircraft and entering air traffic commands to review their acceptability and reliability during DSR's operational test and evaluation period. The HOST and Oceanic Computer System Replacement (HOCSR) was also implemented. This system was need to meet Y2K requirements. HOCSR produces and processes information on aircraft movement throughout domestic and oceanic airspace.

Over the years several corporate downsizings have opened new opportunities and I have acquired systems programming skills. Systems programmers are responsible for troubleshooting system erros that users receive when trying to run jobs. During the DSR and HOCSR projects there were associated support software changes that I had to implement so the new software would run on the old hardware and vice versa. System configuration files were updated and user executable programs were reviewed, tested, and updated as needed.

I am currently in a project manager position; I was responsible for the installation and integration of the projects mentioned above. My primary responsibility is to enhance and maintain the customer's satisfaction with the new computer systems and to manage the updates and enhancements that are needed to correct any reported problems. I have completed graduate courses in education, just in case I decide to leave the corporate world and teach mathematics. With a bachelor's in mathematics you can always build upon that knowledge in whatever field you may pursue!

Yvonne Zhou

BA Mathematics and Computer Science
Macalester College

Software Engineer
Cray Research, Inc.

In May 1991, I earned my Bachelor of Arts degree in mathematics and computer science from Macalester College in Minnesota. Since then, I have been working at Cray Research as a software engineer and systems analyst and am now a graduate student at the University of Minnesota.

Unlike many students, I studied mathematics very willingly throughout high school. I enjoyed working on the problems and did well in class. My grandma always told me, "If you can do well in math, you can do anything." I didn't really believe it, of course. I studied mathematics because I was good at it. For the same reason, I decided to study mathematics as well as computer science when I entered Macalester College in 1987. My favorite subject in mathematics is geometry, and I did my senior-year honors paper on finite geometry. That was something I thought I would never be able to make use of after I finished college, but it later came in handy when I became a graduate student in computer science.

I was hired as a software intern at Cray Research in the summer of 1990 to work on a scheduling program for the Cray Supercomputers. The manager who hired me later told me that the reason he picked me was because I had a very strong background in mathematics. I can't tell you the direct connection between cal-

culus and computer programs, but years of mathematics really exercised my logical thinking. A program is like a math problem in a lot of ways.

Mathematics helped me get my job at Cray, and has also helped me do well on my projects. I have often applied my knowledge of calculus and numerical analysis to the projects on which I am working.

While working at Cray, I am seeking a Master of Science degree at the University of Minnesota. My thesis paper concerns methods that allow faster and easier data access in a database system. I never thought my knowledge of finite geometry would have helped me define the first few sets of methods. During my research, I have discovered papers that define allocation methods using fractals and many other new methods that were introduced by mathematicians.

Mathematics alone may seem powerless sometimes, but when it is combined with another science, nothing else can be more powerful. After working in the industry for over two years, I feel even more strongly about the importance of mathematics to computer science and other related fields.

Yaromyr Andrew Zinkewych

BA Mathematics
Frostburg State College

Scientific Systems Programmer
Bendix Field Engineering Corp Data Capture Facility

Troubleshooter
Hubble Space Telescope

The *Washington Times*, in celebration of the Hubble Space Telescope's second anniversary of exploring the universe, wrote: "NASA scientists say their space telescopes are making a wealth of new astronomical discoveries. These include a black hole at the center of a nearby galaxy and the discovery of what some scientists are calling an entirely new type of quasar." Everyone has heard the jokes about the "near-sighted" Hubble Space Telescope (HST), but the HST has made a number of outstanding observations, and I am very excited to be working on this project.

In December 1979, I graduated from Frostburg State College, a small college in western Maryland, with a degree in mathematics. I majored in math because of my strength in thinking analytically and my ability to do detailed work. These strengths would later prove very beneficial when I joined Bendix. Like many math graduates, I was uncertain of the type of work I could do with my math degree. I applied for many positions after graduation, from actuarial to technological support fields. For several years, I worked as a lab technician, but was unable to develop to my fullest potential at this position.

In 1988, I joined Bendix and was assigned to a NASA contract at the Goddard Space Flight Center in Greenbelt, Maryland. I became part of the Information Processing Division, working for the Data Acquisition and Telemetry Analysis Department (DATAD). DATAD provides telemetry data evaluation, analysis, and accountability for science data received from various satellites. My first project

was the Earth Radiation Budget Satellite, on which I performed telemetry data analysis and aided in troubleshooting data anomalies.

In 1989, I became part of the Analysis group that helped develop the Data Capture Facility for the HST Project, in preparation for its 1990 launch. The Data Capture Facility (DCF) is part of the HST Ground System that captures, processes, and transmits HST science data to the Science Institute at Johns Hopkins University. Our duties included learning to identify, troubleshoot, and resolve problems with receipt of data through the Tracking and Data Relay Satellite System and the processing and transmission of this data through the DCF systems. Also, our group participated in the many pre-launch tests run to validate the HST Ground System.

On April 24, 1990, the HST was launched into orbit by the Space Shuttle Orbiter Discovery. The next day, the DCF received its first science data (and we were ready for it). The Analysis group at the DCF has uncovered and resolved many anomalies that have occurred in the transmission of science data from the spacecraft, in the processing of these data at the DCF, and in the actual science data itself.

In September 1990, I was promoted to Troubleshooter of the DCF Analysis group. My responsibilities include coordinating and assisting in the resolution of major anomalies that affect the HST science data. Also, as part of that position, I became involved in the DCF's support of the HST servicing and repair mission of 1993, in which Shuttle astronauts performed a spacewalk to enhance the HST and compensate for its anomaly.

Paul Zorn

AB Mathematics
Washington University, St. Louis
PhD Mathematics
University of Washington
Professor
St. Olaf College

I teach mostly pure mathematics at a liberal arts college so I might be thought a dreamy, impractical, type. Wrong! Nothing is more "useful" than college mathematics. An analogy with language explains why. Like French, mathematics has lots of uses, but they all require knowledge of the language itself—lots of vocabulary and grammar, and a glance at the literature. Conveying all those things is my job. With a working knowledge of the language of mathematics, one can go anywhere: to concrete applications, to teaching, to more study of mathematics itself.

Since the appearance of the first addition of this book, some things have changed. For instance, the internet would now join digital music and wide-body jets on any short list of popular wonders of science and mathematics. Cryptography—itself driven largely by the growth of the internet—is now a leading application of number theory, a mathematical field traditionally seen as the purest of the pure. (Not all the mathematical news is computer-driven: in the mid-90s, Andrew Wiles solved the 350-year-old problem known as Fermat's Last Theorem.)

Another current of change in mathematics, related to but not entirely driven by computing, has come to be known as "calculus reform." That loaded phrase means quite different things to different people, but many college mathematicians agree that the elementary calculus course, historically the foundation of a mathematics major, needs some shoring up. Too many students fail at or do

poorly in calculus; those who pass may manage some basic manipulations, but still understand little of what calculus is really "about."

What to do about it—how to "reform" calculus—continues to spark "free and frank discussions" (as they say inside the Washington beltway) among college mathematicians. My own opinion is that change is indeed needed. Calculus courses have often suffered from focusing too narrowly on symbol manipulation exercises; at worst, the truly useful, interesting, and important ideas of calculus can disappear in a welter of half-understood formal symbols. A better way, I think, is to combine graphical, numerical, and symbolic approaches, in roughly equal proportions, to offer both variety and perspective to calculus ideas. Modern mathematical computing and graphics, used judiciously, can play an important role in making all of these viewpoints accessible to students.

These opinions, along with the "linguistic" approach to mathematics described above, have led me into mathematics textbook writing. With my St. Olaf colleague Arnie Ostebee I've written a three-volume textbook, *Calculus from Graphical, Numerical, and Symbolic Points of View*, which aims (as the unwieldy title suggests) to combine and compare different viewpoints on the subject.

Is textbook writing a diversion from "real" mathematical work? I think not. For me, writing textbooks raises the type of problems I like best: ones that combine mathematical and expository elements, and lets me use both language and mathematics in search of solutions.

Appendices

Mary Schilling

Seven Steps to Finding a Job

Fortunately, even in times like this, mathematics majors as well as computer science majors are in demand. There will always be a need for entry-level candidates with training, ability, and confidence in mathematics. But that need alone doesn't automatically translate into job offers. The secret to securing a good first job is in the search itself. It's not always the most qualified person who gets hired; it's the candidate with the best job search skills and strategies. So take the time to do it right. Start at the beginning.

1 KNOW YOURSELF

Too many students launch into the job search skipping what may, in fact, be the most crucial step of all: self-assessment. If you ignore this stage of the search, you proceed at your own risk.

In a survey conducted by Northwestern University, 500 employers were asked to note job applicant behaviors, responses and activities which were counterproductive to the job search. Among the top weaknesses listed was applicants not knowing themselves. Particularly at entry-level, self-assessment encourages the job applicant to engage in a systematic evaluation of interests, skills, attitudes, and values.

You can choose to use a computer-assisted program in your college career center (DISCOVER and SIGIPLUS are popular) or a paper and pencil assessment survey (the Campbell Interest and Skill Survey, the Holland Self-Directed Search, and numerous others), but spend some time identifying what it is you most enjoy doing, what you do best, what's important to you, and how you feel about various kinds of work functions and environments. Figure out what you have to offer and what factors are important to you for job satisfaction, not just as a math major but as a whole personality. Not only will this analysis provide direction for the remainder of the job search, it will also prepare you for that somewhat intimidating and often-used first interview question: "So, tell me about yourself."

Mary Schilling is the Director of the Career Development Center at Denison University in Granville, Ohio.

Identify the skills you have developed both through your academic and co-curricular experiences. It is here that math and computer science majors and minors have a decided advantage. Employers are actively seeking women and men who have developed quantitative reasoning, analytical thinking, and problem-solving skills. You know how to manipulate and crunch numbers as well as to conceptualize and theorize. Your discipline has trained you to think logically and to attend to detail. You can deal well with abstractions; and you can apply theory to practical problems. You know the language of both business and the sciences. While these skills are among the most highly valued by employers, they are not enough.

Today's employers are looking for mathematically inclined women and men who also have developed excellent skills in oral and written communication and effective interpersonal skills. Successful job candidates have proven they can work on a team, organize projects, and deal with stress. Employers look for candidates who strive for excellence in all aspects of their lives. Spend some time carefully considering your interests, attitudes and values. This will help you to determine where and how you want to use your abilities.

2 SET YOUR GOALS

Once you've articulated the skills you have to offer, it's time to move ahead in your search and set some specific goals. You will need to determine the functional areas and the environment within which you want to use your skills as well as the geographic area.

Even in a tight job market, math and computer science majors are competitive within a broad—almost limitless—range of job areas. Private business and industry, the public sector (local, state, federal government), education, and non-profit organizations all provide opportunities for candidates with a strong mathematics background. Positions are available in underwriting, banking, investments, marketing, research and development, programming, systems, consulting, actuarial, and a myriad of other arenas. The secret is to research those job options which appeal to you and to focus on several which match what you've learned about yourself through self-assessment. College career centers typically offer library resources about career possibilities for the math and computer science major. Public libraries as well may have career resources for your use. Hopefully, an internship or a co-op experience has already given you the kind of information you need to narrow your search to a manageable number of relevant opportunities.

3 PREPARE A POWERFUL RESUME

Since employers receive numerous resumes for each position and may spend only 30–45 seconds glancing at each one, it's extremely important that your resume be quickly and easily digested, attractive and powerful. While standard resume advice is readily available in guides, a few hints might be helpful to the math major. If you are clearly focused on one or even a couple of functional areas or job titles, write a crisp objective which articulates the kind of position you hope to attain. Be sure to emphasize what it is you have to offer the employer, not what you want for yourself. If you are not yet focused, don't include an ambiguous or multidirectional objective. Instead, articulate your objective for a particular position in the accompanying cover letter. Or consider the possibility of using a summary of qualifications rather than an objective, making sure in either case that you emphasize your skills in quantitative reasoning, analytical thinking, and problem-solving.

Within the section of your resume describing your educational credentials, list "relevant coursework," including titles of math and computer science courses you've taken. Or include a section on "computer expertise," listing the various languages you've learned whether in the classroom or on the job. Be sure to include senior research or other research experience. In listing your internship, summer job, or part-time work experience, tell the reader what you've learned (including computer software programs), not just what you've done. Show evidence of your skills, noting specific achievements and accomplishments both in your experience and your activities sections. Use quantitative measures of your achievements whenever possible. Remember, the employer is trying to determine if you are a good match for the position and needs to be convinced that your past experience, both educational and work, is relevant and a good predictor of your potential.

Now that you have a strong tool for marketing yourself, what do you do with it? Gone are the days when sending out 300 copies of an entry-level resume to a list of potential employers resulted in a good response for follow up. It's just not that easy. You need to establish a "hit list" of employers, based on your self-assessment and your focus on particular jobs, environments, and geographic areas. To create such a list, you will need to continue your research and begin to develop a network.

4 ESTABLISH A NETWORK

Though already overworked to the point of becoming a buzzword, "networking" is an important and effective tactic. It's been said, "If you don't network, you may not work."

While some students continue to find success in on-campus recruitment, many more secure their first position through a series of connections with college alumni, faculty, family, friends (and friends of the family), acquaintances, community contacts, and previous internship or job supervisors. These connections can be used to establish informational interviews, access to good career advice, general support through the job search process and even actual job interviews. Just remember that these contacts are volunteering assistance; don't take them for granted or act as though you are entitled to their help. Send a resume and a letter in advance of your meeting or phone conversation. Be prepared for the interaction with a clear focus, articulate questions, a specific request for advice, job leads, information or names of other professionals in their network whom you might also contact. Be persistent in pursuing network leads. And, finally, follow up with thank you letters or friendly phone calls to keep your network contact informed about your progress.

5 APPLY FOR POSITIONS

To launch the actual application process, you will want to get job listings from a variety of sources. Your college career center or placement office will offer on-campus recruitment programs as well as job listings. Professional mathematics journals or magazines may have appropriate postings. While newspaper ads are limited in their helpfulness, don't ignore them. Participating in a regional or national data base and attending job fairs in the area of your geographic focus may also be source for job leads. Your network contacts are also likely to produce information about openings.

Prepare a specialized cover letter for each position, highlighting specific aspects of your educational, work or cocurricular experience relevant to the job. Indicate in your letter how references may be secured. Also, be sure to mention when you are available for an interview. Unless directed otherwise, indicate that you will be following up with a phone call; keep "the ball in your court."

Keep copies of all of your job search correspondence. Systematically follow up each job application with a telephone call to confirm the receipt of your letter and resume and to inquire about the prospects of a job interview. The purpose of your phone call should be to show an interest in the job opportunity, to gain information regarding the timeline of the job search and to assess your prospects for an interview. While it's good to be persistent, be careful not to seem demanding or overbearing. Above all, be polite and professional.

6 PREPARING FOR THE JOB INTERVIEW

Having achieved your goal of attaining an interview, now you must prepare for it. Do your homework on the organization or company with which you are interviewing; if you don't, it will show. (Lack of knowledge of the employer was a second glaring weakness noted by the interviewers participating in the Northwestern University survey.) Use your career library and the public library, check industry directories, talk to folks in your network, read magazines and journals to research the employer. Again, there are more interview guides available than you will ever need, but find a good one and review your basic interviewing skills or attend an interviewing workshop offered by your career center.

Anticipate questions; many are predictable. Formulate possible answers; practice with a friend. The trend in interviewing today is toward situational interview questions. Well-trained interviewers will ask you for examples of situations in which you used your skills. You'll be asked to describe a situation, the action you took to address it and the results. Choose examples from your academic work, from internships or summer jobs, or campus activities. Have ready descriptions of situations in which your quantitative skills and problem-solving abilities were particularly effective, for examples, tutoring other students or participating in mathematics competitions. Interviewers are looking for self-confidence, enthusiasm, good communication skills, career direction and flexibility. Be ready with evidence of your teamwork, your dependability, your perseverance, your eagerness and ability to learn, and your willingness to work hard. You need to convince the interviewer that what you offer and what the employer needs are a good match.

Have good questions ready to ask the interviewer, indicating a sincere interest in the employer and an awareness of the employer's needs. Ask about the growth of the organization, new products/technologies, company culture, management style, career paths and opportunities for advancement, and training programs. You are being judged by the quality and content of the questions you ask the interviewer as well as by your responses you give. Save questions about salary and benefits for the second or on-site interview.

Finally, convey to the interviewer that you really want the job, that you would be excited about working for the employer, and that you look forward to talking with them more about the opportunity. Within a day or two follow up with a thank you letter, reminding the interviewer of something specific about you and your conversation and reiterating your continued interest in the job prospect.

7 EVALUATE OFFERS

Once you are extended job offers, evaluate them based on how well they match your interests, skills, values, and attitudes. Determine if the environment is one in which you can do your best work. Consider the opportunities for professional growth the positions may afford you. Don't hold out for the perfect job. There are none! Since a first job is not forever, think about whether or not the position will provide a good foundation, a launching pad for your career. In evaluating geographic location of job offers, remember that employees who are open to relocation early in their career are highly valued and will have the benefit of geographic stability in later career stages.

In evaluating salary offers, be realistic. The July 1993 College Placement Council Salary Survey reports salary offers extended to 1993 graduates at 402 career centers at colleges and universities across the country. Offers to Bachelor's Degree Candidates in mathematics, including statistics, averaged $26,000. While offers to candidates with Bachelor's in Computer Science averaged $30,900. What quoted salaries don't reveal are benefits, profit sharing options, bonuses, commissions, relocation reimbursement, tuition reimbursement for graduate or professional school courses, among others. Once offers are extended, you will be provided with details on the entire remuneration package, which may provide a different picture than salary alone. Complete your job search with courteous follow-up correspondence to all who helped you. Express your appreciation for assistance and support. You may want to stay in touch with some of your contacts. And once you have established yourself in your own career, pay back the system by offering to include other young jobseekers in your own growing network.

A Boeing advertisement in the College Placement Council Annual magazine boasts: "You can go as far as you want … It's a matter of discipline and degree." As a math major, you're in a good position. Your discipline and your degree provide you with a set of skills which are in demand, skills that are transferable and flexible even in a challenging job market. The opportunities are there, waiting for you. Take your job search seriously, work on it systematically, and remain confident. You'll be pleased with the results.

Mary Schilling

Interviewing Tips From the Pros

You spent hours (days?) perfecting your one-page resume. You circulated it within your network and through a structured on-campus recruitment program. Fortunately, it has been successful in securing interviews. Now what?

Interview Preparation

Career advisors say that you need to do only three things in advance of your interviews: Prepare. Prepare. And prepare some more. Interview preparation is not like cramming for an exam; it's an extended process. Your preparation began long ago and has been developing as you have made important choices: college, major/minor/concentration, co-curricular activities, internships, summer jobs, and work study. You started preparing specifically when you began working on your resume. With interviews now scheduled, there are only two more things you need to know.

First, know yourself—your abilities, attributes, values and attitudes. Assess your skills and determine specifically what you have to offer the employer. As a mathematics major, consider your skills in quantitative and analytical reasoning and how they helped you to become an effective problem- solver. Prepare examples illustrating each of the various skills you want to market.

Second, know the employer. Research the industry in general and the employer in particular. You'll find literature and resources in your career development center as well as in college and public libraries. Request the employer to send corporate/organizational literature. Network with current employees (especially alumni of your institution) as well as your professors. Arrange information interviews by telephone or in person. Learn as much as you can about the employer's corporate philosophy, structure and culture, the recruitment process, the position description, and, most importantly, the skills and abilities they are looking for. Assess the ways in which the employer could benefit from your math-related skills. When adequately prepared, you can divert the adrenaline that comes with nervousness into a positive energy that makes you alert and "pumped" for the interview. Now, you are ready for the "face-to-face" interview.

Salable success factors valued by the employer

Prepare responses, noting your Assignment, Action, and Accomplishment

Money—Think of a time you saved or made money for a campus organization or summer employer.

Time—Describe an action you took that increased productivity or saved time for an internship or co-op employer.

Efficiency—Can you think of a problem you solved speedily, logically, and accurately? Did your math major help?

Organization—What event, activity or project have you planned and implemented from beginning to end?

Teamwork—Were you ever involved with any team projects, sports, or activities? How did you and your team work to solve a particular problem?

Hiring or recruiting—Have you ever hired people or recruited volunteers? What skills did you find helpful?

Risk-Taking—What was the last "risky" situation you were involved in?

Public Speaking—Did you ever have occasion to speak in public? How did you prepare yourself?

Adaptability—Describe a situation when you were called on to be flexible or adapt to a new situation.

Helping others—Think of a time when you helped someone in your community, on campus, in your family.

Perseverance— Describe a time when you had to handle challenges and obstacles to complete a particularly difficult task or assignment.

Innovation—Have you ever come up with a new idea for an organization or summer employer?

Making Improvements—Have you ever observed the way something was done and figured out a better way to do it?

Beginning the Interview

Much has been written about the first few minutes (would you believe 30 seconds?) of the interview. Some assert that the initial impact of appearance, the greeting and hand shake, and the conversational style in this "warm-up" period can result in a decision to offer a position. Others argue that offers are not won by these initial impressions, but they certainly can be lost. Whatever the case, remember: You never have a second chance to make a first impression.

Use "warm-up" conversation to build rapport with the interviewer. Whether it is about your college, the weather, sports, or your hometown, it's meant to put you at ease, to help you settle into the interview. How well you handle this level of conversation is important, since you will have many opportunities for "small talk" in dealing with the employer's clients or customers.

Don't underestimate the power of non-verbal communication. Although word choice is critical in delivering the message, voice tone and body language have a greater impact. Facial expressions, posture, gestures, and general appearance play

Frequently Asked Interview Questions
(and the qualities they seek to identify)

Under each quality is a series of questions that aim to give the employer an idea of how you rank in that category.

Intelligence and analytical ability

What were your favorite and least favorite courses in college?

Would you describe yourself as more analytical or more verbal?

Does you grade point average reflect your academic ability?

Why did you choose to major in mathematics?

Work experience and required technical skills

Can you walk me through your resume?

What specific skills did you develop in your internships or co–op experiences that you would consider useful here?

Leadership qualities/ team–playing ability

What makes you better than everyone else I'm interviewing today?

Why should I hire you?

Tell me about your extracurricular activities?

Energy and stamina

Describe your average weekday.

How do you feel about sixty-hour work-weeks?

Initiative and entrepreneurship

What one thing would you change about your college?

Describe one change you instituted at your last summer job?

Where do you see yourself in five, ten, or twenty years?

Maturity

What are your career plans?

How did you choose your college?

Tell me about one of your weaknesses.

Communication Skills

Tell me about your senior thesis.

Describe your responsibilities on your last job.

Read any good books lately?

Creativity and flexibility

Why are manhole covers round?

You can have dinner with anyone through-out history. Whom would you choose?

Interest in the position

What fields are you investigating for em-ployment?

What other firms are you interviewing with?

Strictly hypothetically, of course, if we were to give you an offer today, would you accept it?

Describe your ideal job.

What is your current understanding of what we do here?

What do you see as the greatest problem facing our industry right now?

Personal qualities and personality

What do you do for fun?

Describe your ideal weekend.

How do you choose your friends?

What's the funniest thing that's ever hap-pened to you?

an amazingly important role in determining the impression you make. Participate in a mock interview so you can see yourself on video as others see you. Some of your communication habits may need correcting or improving. You can assess the total impact of your verbal and non-verbal communication style.

Marketing Yourself

It's confidence-building to think of the interview as a two-way process. Both you and the interviewer are shopping; both are marketing. The interviewer is assessing whether you are the best candidate for the position, one who offers something that others don't, one who can do the job, fit into the culture of the workplace, and make a difference. At the same time, you are trying to decide whether or not you would really like to work for this employer, whether the job opportunity is well-matched to your interests, skills, and values. In effect, you are interviewing each other.

You are marketing yourself as a "product" which will meet the employer's needs. You want to help the interviewer know your "features" but, more importantly, the benefits you can bring to the employer. Dorothy Leeds calls this "benefit selling." It assumes "the only reason an employer will hire you for a particular job is because there's something in it for him."

Throughout the interview, you will have the opportunity to emphasize your features and benefits. What features you choose to highlight and how you customize your potential benefits will depend on your research of the employer.

While nearly all interview guides deal with transferable skills, Leeds deals with "salable success factors" and encourages you to develop a structured way of responding to the interviewer's questions. She talks about the AAA's of selling yourself: **Assignment**—a particular situation you faced, **Action—**the strategy you employed, and **Accomplishment** —the results of your action.

The assumption that past performance predicts future performance is used extensively by the best-trained interviewers. It's effective in getting the information the employer needs to assess the transferability of your skills; it's also an excellent way for you to showcase your experiences and relate them to the potential job.

Remembering that your features remain the same for all employers, choose benefits that fit the particular job. In your responses emphasize the abilities that will be of value to the particular employer. Emphasize results. Leeds recommends that you keep 10 "benefit statements" in mind so you can choose one when needed. Have a minimum of three examples of objective/strategy/results "stories" to demonstrate each relevant skill.

Think through and prepare these responses. Memorize the content and the theme of each; so you will be ready to improvise depending on the interviewer's question. Explore different approaches to these questions; talk them out with a friend; stand in front of a mirror. Prepare until your responses flow naturally.

Most Often Asked Questions

While you can find plenty of lists of "the most often asked interview questions," what you need to figure out is why the interviewer is asking each question. In *Hot Tips, Sneaky Tricks & Last-Ditch Tactics,* Jeff Speck organizes the intent of the most frequently asked questions into categories. [See the box on page 231 for his categories and questions. Make sure you structure possible answers for each before the interview.] Once you know the category and the skills being sought, you'll know how to respond or, to put it bluntly, "to give the interviewer what he wants." The interviewer will rate you on each of these categories; your answers will determine your ranking.

Imagine other questions and what the employer is trying to find out about you. Expect the unexpected. While most interviews are not designed to be stressful, occasionally an interviewer will "throw you a curve" to see how well you respond.

Your Turn to Ask Questions

Some interviewers encourage you to ask questions. They may provide time near the close of the interview for this. This is at one level a courtesy, but also interviewers see the kind of questions you ask and how you ask them as another way to judge your candidacy.

The person asking questions tends to be in control of the interview. By asking well thought out questions about the employment opportunity, you will control the conversation and can steer it in the direction you want. The more you ask, the more control you gain.

Organize your questions in advance. You may want to have a copy of them with you for reference in case you have forgotten some questions. This lets the employer know you have given considerable thought to the interview and are prepared.

Use this portion of the interview to ask open-ended, neutrally phrased questions. Make your questions relevant to the position for which you are interviewing and specific to the employer's needs. This is not the time for asking "yes" or "no" questions. You can also ask questions about the interviewer's experiences with the employer. Asking about something positive you've recently learned about the employer is a good way to end this portion of the interview.

Prepare a question for each attribute you want the interviewer to know about. If certain attributes were not discussed, ask the related question to let you reinforce a key strength. You'll find a list of suggested questions to ask the employer in nearly every interview guide. Thoughtful questions might include such topics as:

- financial stability/growth of the organization
- new products/technology
- training program
- supervision/performance reviews
- travel/relocation
- typical day/first year assignments
- company culture/management style
- career paths/opportunities for advancement
- need for graduate/professional school degree

Closing the Interview

Recruiter feedback indicates that a major weakness of students is failure to ask for the job. Don't just let the interview end; take the initiative and close it as a salesperson would "close the sale." No matter how great you look, how firm your handshake, or how impressive your answers, if you can't convince the interviewer that you really want the job and why you want it, don't expect another interview, much less an offer.

You will need to find a way to express your genuine interest. Try something as simple as: "I'd very much like to work for your company. When can I expect to hear from you?" An expression of sincere interest in the position may keep you in the running for a job offer.

The Follow-Up

While many of today's employers don't consider thank-you notes necessary, it remains both a courtesy and a smart strategy to send a letter following your interview. You should follow two rules.

First, send your typewritten thank-you letter, the same or next day. True, the decision to offer another interview (or even the job) may have already been made, but the promptness of the follow-up suggests an energy, a professionalism, and a thoroughness that are valued by employers. If you are on the border line, it may be a decisive factor; if you are invited back for the next round, your immediate follow-up becomes a plus in your candidacy.

Second, make your thank-you letter distinctive and specific to your interview. Reference an attribute which you have and the employer needs. Remind the interviewer of some "story" you told about how your response to a particular situation had outstanding results. Tell the employer how helpful the interview was in increasing your understanding of what the organization does and what kinds of career opportunities it offers. Finally, reiterate your interest in working for the employer and the fact that you are looking forward to continuing in the recruitment process and to visiting the company and meeting other employees. If appropriate, suggest that you will be following up with a telephone call within an earlier agreed upon time period.

Because you have met the interviewer, the tone of your letter can be a bit more casual and personal. But remember, the purpose of the letter remains professional, and the letter may be shared with others. If you are fortunate, it might become a part of your employment file!

Many successfully employed people will tell you "it's not what you know but whom you know." Or it's all about "being in the right place at the right time." Or that it's a matter of luck. Don't believe them, unless, of course, you accept Thomas Jefferson's definition: "Luck is when hard work meets opportunity."

In that spirit, good luck in interviewing!

References

Dorothy Leeds, *Marketing Yourself: The Ultimate Job Seekers Guide*, Harper Collins, 1991.

Jeff B. Speck, *Hot Tips, Sneaky Tricks & Last-Ditch Tactics: An Insider's Guide to Getting Your First Corporate Job*, John Wiley & Sons, 1989.

STUDENTS SPEAK OUT

Math Majors Tell (Almost) All

The students interviewed in this column all attended the Joint AMS/MAA Mathematics Meetings in Cincinnati in January. MAA Visiting Mathematicians ***Fred Rickey*** *(Bowling Green State University) and* ***Anita Solow*** *(Grinnell College) chatted with them in the Student Hospitality Room about two topics: the advice they would give to students who had an interest in mathematics and why they chose to be mathematics majors. What follows is an edited version of their remarks.*

What advice do you have for students in mathematics?

Rebecca Field
Bowdoin College
Sometimes if a person doesn't have a particular faculty member take an interest in them, they will accidentally slip through the cracks, even in a small department. If you like the subject of a class, go talk to the professor about it. Make sure everyone knows you are interested in mathematics. Once they know, they will do anything for you. Unless you go up to a math faculty person and say "I am interested in working in mathematics for the summer, is there anything you know about this?" you will not find out. Some programs, such as REUs, are hard to find out about. [REUs are Research Experiences for Undergraduates].

Sue Sierra
University of Michigan
The biggest thing is to be open minded—don't rule out things too soon. If you are going on, make sure people know about it. It is harder to get people in your department to take you seriously if you are Black or a woman. You may have to get the information yourself. I did a summer program at Mt. Holyoke. I did not necessarily get any research done, but it was a fantastic experience.

Bill Correll, Jr.
Denison University
I worked at the University of Dayton last summer. That experience put a lot of things into perspective for me. The opportunity to work with some really talented and dedicated peers and mathematicians was invaluable.

Rebecca: Everyone should consider a summer research program. You also get paid. It's really necessary since most of us need a job for the summer. They are wonderful experiences, even if the research itself doesn't work out. Just knowing other people who are serious about mathematics is so important. The programs are really competitive, so apply early. Also get involved in programs at your own school.

Melonie Gordon
Francis Marion University
I agree that you should definitely apply to summer math programs. I was involved with the Berkeley Summer Math Institute, and this was a wonderful program. Speakers came in and let us know about math that we didn't know existed and about different career opportunities. We had two math classes, and we did math all day, pure math and applied math. It gave us a chance to experience and see if that's what we wanted to undertake. I would suggest that if you are considering math, you try to find a program to see if that is what you really want to do.

Tanya Henneman
Spelman College
The first summer program I did was a joint program between Bryn Mawr and Spelman College. We broke up into small groups and did research on wavelets or graph theory. This past summer I also was at Berkeley. I took two courses, Calculus of Manifolds and Numerical Analysis. Lots of invited lecturers came and talked to us about their math research. We also had workshops on how to give math presentations, careers, and various things that would help us in graduate schools. It gives you a good idea what grad school would be like. That's how they treated us—like graduate students. It's not a laid back summer, necessarily. You learn a lot, and they really expect a lot. It is very demanding. It just helped me all the more for my final year. I felt so much more prepared. We still keep in touch. It is great seeing each other here at the conference.

Shurron Farmer
Florida A&M University
I also particpated in the Summer Math Institute at Berkeley. It was a very good experience. I learned new ways of looking at mathematics. It was good to see a program where minorities were highly represented.

Grady Bullington
Western Kentucky University
I did things differently. I completed my undergraduate degree in three years, with a double major in computer science. Advice I would give: Enjoy your four- year program and take advantage of the time you have. Maturing in mathematics takes time. Don't be afraid to feel stupid, because we all feel stupid sometimes. Even our professors feel stupid when they are doing research.

Melonie: Don't become discouraged once you get into your upper-level math courses where you start trying to prove things. Talk to your professors and really work at it and don't stop just because you become frustrated, because you will become frustrated.

Tracy Lilly
Austin Peay State University
You need to get to know, not only the professors of the courses you are taking, but all the professors in the department. I think that helps to give you a broader scope of the subject you are interested in.

Laura Harmon
Hiram College
I think working with other students helps a lot—getting other people's ideas.

Shurron: Learn the importance of study groups. That's my main advice. Throughout all of my undergraduate years in both biology and mathematics, study groups and peer mentors were highly stressed. And it was one of the things that I loved about the Berkeley program. It is essential in graduate school.

Melonie: Communicating your ideas to your group helps you understand the concepts. It is so important to learn to communicate mathematical ideas, especially if you want to go on in mathematics.
Tanya: Doing research and independent study are probably the best ways to test what you know and don't know.

Grady: I wish I had spent more time in the trenches of the book—two hours for every hour of lecture. The more math I do, the more I want to do.

Tanya: Don't treat your courses as separate courses. Treat them as vital even if they are freshman courses. A lot of times students think that Linear Algebra, Calculus, and Real Analysis are all separate courses. They may have separate titles, but in a lot of ways these courses are interrelated. You are going to need to know and understand the courses that you took during your freshman and sophomore years because they will come back eventually when you take Analysis or Abstract Algebra, and your profs expect you to know and understand those calculus and linear algebra concepts. Treat those classes seriously and really try to understand them, because they provide a basis to build on. If you don't have that then you will keep having trouble in your later classes.

Melonie: Another thing I learned to do was to read mathematics. Early on I was doing the problems, but now it is much more helpful to actually read the text that comes before the problems to get a better understanding about what is going on. At Berkeley, we had to read our math texts before doing the problems, and now I am doing a much better job at reading.

Tanya: We had this one text, a very small book, and when you have a book filled with 'clearly' and 'obviously this shows,' you have to learn how to read it because it is not clear and it is not obvious. A lot of times we spent a couple of pages proving what was claimed to be obvious. It was good training.

Bill: My last bit of advice is to work lots of non-routine problems. I am a big problem solver; I just like to do any old problem I can get my hands on. If you just get stuck in one kind of math, that can hurt. Keep your skills sharp.

Erik Burd
Sonoma State University
An excellent idea for students is to enroll in math competitions, such as the Putnam and the Mathematical Competition in Modeling (MCM). Over the past four years I have really been engrossed in it. It has increased my interest in math and made me more aware of its importance, because I was using my math skills to solve a problem that is applied to industry. I thought it was a good learning experience and a lot of fun.

Brian Bush
Hiram College
This will be my third year in the MCM, and it has really helped out a lot. The first year we did it, we used computer programs, calculus, and whatever we could find over the whole weekend. It really gives you insight into what people do with math.

Melonie: The MCM competitions are really exciting—working with the math, going to the library. That whole weekend was just fun filled with math.

Jennifer Courter
California Polytechnic State University, San Luis Obispo
I am glad that I took the Putnam exam and participated in the Modeling competition. One thing I wish I had done more is look ahead. At the beginning of my undergraduate career I had an idea that I might want to go to graduate school, but now I wish I had done more to prepare for applying to graduate schools. I am glad that I have gone to the AMS/MAA meetings. Attending the meetings has really changed my perspective, because I have seen the mathematical community as a whole and what is going on outside of my course work. I have learned about possibilities for the future. So, look ahead. Some people may have trouble getting funding to travel, but everyone can benefit from attending the math meetings. You even have the opportunity to meet students from other universities.

Laura: You get to meet other professors, too.

Brian: It's kind of good to see what actually happens when our profs take off to go to a conference. What do they actually do there? It's good to see other profs getting together and sharing ideas and talking about their area.

Jennifer: You get exposed to areas of mathematics you never knew existed.

Tracy Lilly: I found that there is a lot of information on different careers in mathematics at the meetings. That really helps out whether you have decided on a career or if you are still out there searching.

Melonie: Another thing that keeps me motivated is going to conferences like this and talking with other math majors. It seems like we all have the same ideas and conversations, and when I go back home I think about all of you facing the same problems. And that helps a lot.

Laura: You know, another thing is that I wish I had taken more com-puter courses.

Brian: I agree. I'm actually a dual major in math and computer science. Computer science has helped me a lot. It taught me to think through the steps of a process, which made the mathematics better.

Erik: I always kept my computer science background as a foundation for my math. It does help to explain the principles and concepts in mathematics, because it allows you to visualize them. I would definitely recommend at least four courses in computer science for everybody.

Shurron: One piece of advice that I would give to other undergraduate mathematics majors is to develop an appreciation, if not a love, for computers. I don't love them, but I do appreciate them. They are becoming more and more useful, actually more and more essential, in some aspects of mathematical research.

Sue: I have some more advice. One thing to realize is that doing math in grad school doesn't necessarily mean you'll get a job after you get your doctorate. I have a lot of friends who are just about to get their degrees and are looking for jobs and are having a really hard time. The job market is really bad right now and isn't expected to pick up until 2010 or so. I'm still doing mathematics, but you should be aware that you could go to graduate school, work really hard for five or six years, and be unemployed on the far side.

Sue: And don't forget to learn how to cook before you move out of your parents' house, because you won't have a chance afterwards.

Rebecca: And try to take a pot or two.

What made you decide to be a mathematics major?

Brian: I took the easy way out. I can't write anything so English is definitely out, so I became a math major. Then I had to write proofs and there went that theory.

Tanya: I came in as a psychology major. In high school I did pretty well in math, but I never thought of it as a career. I took calculus my freshman year, and I just really enjoyed it. My math teacher asked me why I was a psychology major, and said that I should be a math or engineering major. I took a couple of more math courses, and enjoyed them as well, so I decided to change my major to math.

Jennifer: I started out in college as a landscape architecture major. I was taking calculus as an elective, and I liked it a lot. I thought anyone who takes calculus for fun should consider mathematics as a major. I switched to mathematics the next term. Another factor in switching to math was the teachers I have had, especially my 8th grade math teacher.

Shurron: I always knew that I would be some type of science major. I entered as a biology student, but I decided at the end of my first year that instead of being an MD, I wanted to be a PhD. At the same time I was wondering if I could double major in math and biology. The program that I was in, run by the Office of Naval Research, stressed the need for minorities to obtain doctoral degrees. But it was not until the second semester of my sophomore year that I took calculus as an elective. And I did very well and I enjoyed it so much, and I decided that now that I know that I want to teach, I finally know that I want to teach math. So I went ahead and changed to math.

Jonica Harris *Spelman College* You don't think of math as a career. When I tell people that I am a math major they say , "So you are going to be a teacher," and I get really tired of that even though I want to be a teacher. I think there are lots of other things you can do with a math degree, and we need to work harder to communicate that at the high school level.

Melonie: I came in as a math major. I always loved math when I was in school. I liked the hands-on activities. Math seemed to be the hands-on homework. You got your pencil and your paper, and you are always thinking and working things out. If they didn't work out you tried something else. When I came in that's what I wanted to do. I liked working with mathematics, so I came in as a math major with a computer science minor.

Jonica: Well, in high school, what motivated me to be a math teacher were my teachers. They made it fun. And I was the top person in class, which made it better. I was also in an education program in high school and I had my own 9th grade class, 5th grade class and kindergarten that I was teaching in. So now I am in math education and want to teach eventually.

Tracy Lawrence *Spelman College* I came in as a biology major, and then after my freshman year, after I had taken calculus and I really liked it, and I didn't like biology so much, I switched to math/pre-med. I wasn't exactly sure I wanted to go to medical school, and I knew I didn't want to be a biologist. I am now planning on going to medical school.

Erik: I started as a computer science major. I originally was not very motivated going into mathematics, because there appeared to be an over-emphasis on teaching at that time. I didn't realize the importance of mathematics in a variety of subjects until I took a course in Math Modeling in 1990. I was amazed by the large number of applications in biology, computers, physics, statistics, and much more. I was enlightened so much by this that I decided to major in Applied Math. This was the best decision that I ever made in college.

Tanya: In high school, another reason I never considered math as a career was that I always had white male teachers for my math courses. I can only remember one female teacher. I never thought of it as a career for me, because I never saw anyone who was like me teaching math or doing things in math. I think that's one of the great things about Spelman. Everyone around me looks like me, and so I have no reason to believe that this isn't a career that I can do.

Jonica: I had the opposite experience. I had all female teachers and one male. The male teacher made it most interesting. He was kind of biased, so when he saw a Black female that was doing so well, he kind of pushed. He was surprised so he pushed and made it better.

Sue: I kind of fell into being a math major by luck. I had no idea what I would major in, but I was pretty sure that it wouldn't be mathematics. I took a math class my first semester, and it was amazing. I always liked mathematics, but this one discrete math class—every undergraduate's favorite— was a really neat class where we did various topics that are easily accessible to someone who does not know much mathematics. After that I took all the math classes I could. That worked well for me, but not everyone gets that intense about it.

Tracy Lilly: It was my high school mathematics teacher. It was his teaching style that made me interested in math. He made me appreciate mathematics and learn how to apply it and enjoy it.

Bill: I liked Chemistry and Physics and Spanish a lot in high school, but not nearly as much as I liked math. My parents didn't push math on me, but they encouraged my sister and me to develop our math skills. Around sixth grade, I took my first math contest. What fun! During junior high and high school I signed up for many more contests, and tried to prepare myself for them. They laid a firm foundation for further study and helped get me some scholarships. A lot of independent work and highly motivating professors convinced me that I had made the right decision.

Grady: The reason I became a math major was to supplement my computer science major and to give me a broader range of job choices. At the end of my undergraduate career, I realized that I wanted to teach and do research in mathematics, so I enrolled in the Mathematics masters program. Now, I want to work toward a PhD.

Rebecca: I've always wanted to do mathematics.

Anita Solow

How to Really Get a Job

In the first two issues of Math Horizons, there was a pair of articles about how to get a job written by Mary Schilling, Director of the Career Development Center at Denison University. In this issue, we go to the source itself-—recent mathematics majors who got jobs. These math majors graduated from a variety of colleges and universities in the past few years and are now employed in vastly different occupations. They all had different experiences in the job market, but they all have one thing in common: they were successful. Hopefully their stories and the advice they give will help those of you who will be seeking employment soon.

In order to find the people to interview, I called over 30 schools to get the names and phone numbers of former majors who are now employed. Most of the schools were able to supply me with some leads, although several admitted that they do not keep track of their former majors the way they keep track of their graduate students. I conducted these interviews over the phone.

What struck me about these interviews was the number of times that it was mentioned that majoring in mathematics demonstrates to an employer that the person is not afraid of intellectual challenges. Employers seem to respect mathematics majors, not for what they have learned, but for the thinking skills that they have acquired. It was also stated several times that majoring in mathematics provides solid analytical training which enables a person to learn what they need on the job. So, the employers perceptions of the value of majoring in mathematics is well founded!

Most of the mathematics majors interviewed did not have many ideas about what careers to pursue. They knew what they didn't want to do, but weren't aware, at first, of the possibilities open to them. Perhaps the most important step in a successful job search is to get information about careers in mathematics.

Deborah Naftoly graduated from Lafayette College in 1993 with a joint mathematics/economics major. She is currently employed as a staff con sultant for Anderson Consulting. She was hired for her job which requires a great deal of programming in C even though she had never programmed before. Deborah's story is fairly typical of mathematics majors. She did not know what she wanted to do with her major and thought that a math major was only good for teaching or graduate school. After going to the campus career center, she sent out resumes to companies coming to campus to interview. Before she interviewed with Anderson, she spoke with a friend of a friend at Anderson to find out what they wanted in an employee. She also read up on the company. During the interview, she made it clear that this was the job she wanted although this was her first interview.

When I asked her what qualifications she had that she felt made her attractive to her employer, Deborah listed the leadership positions she had at school and her organizational skills. She had supported herself throughout college and felt that this demonstrated that she was capable of taking on responsibility. Deborah felt that her technical background was not as important as her ability to learn. She was not afraid to take on the job.

Her advice—stress your analytical ability. You can pick up knowledge as you go along as long as you are able to think and learn. Keep your options open and stress your positive attributes.

Jeffrey Schneider is currently a high school mathematics teacher in Minnesota. He graduated from St. Olaf College in 1993. Jeffrey knew that he wanted to be a high school teacher, but he still had to work hard to find a suitable position. He sent out over 60 applications to schools in the Minnesota and Wisconsin area and had seven interviews.

Jeffrey's enthusiasm for teaching came through in our interview, and he felt that it helped him land his job. He was also willing and eager to teach lower ability algebra and to make that course as exciting as possible. He was up front and honest during his interview and admitted what he did not know and did not try to pull the wool over their eyes.

He advises mathematics majors seeking jobs not to limit themselves to one particular geographic area. Be willing to take chances and apply everywhere. After all, you can move in a few years if you want to.

Jeanette Elias was a mathematics/physics double major at Agnes Scott College. She graduated in 1992 and is currently at the First Union Bank. While a senior, Jeanette decided that she wanted a career in the financial industry. To find a job, she did everything—newspapers, career office, putting the word out with people she knew. She found her first job in the newspaper. She researched the company as well as she could, but found it difficult to find out much information about a small company. The person interviewing her was a good salesman, and she accepted an offer. However, after three months, Jeanette found the company was not reputable, and she knew that she had made a mistake and did not have the job she wanted.

By this time she had gotten her license as a stockbroker, she was reading the industry magazines, and she had contacts. She heard about a job from someone she knew professionally. She researched this new company both through formal sources at the library and informal sources. She asked friends and relatives if they would do business with this company. When she called about the job, she was told it was no longer open. However, two weeks later, they called back and invited her for an interview. She then was offered the job.

Jeanette found that employers were impressed by her major. Most of the other applicants for these jobs were in economics or business, and she felt that her mathematics and physics background was a big advantage. It put her on an "instant pedestal." She learned how to think in school, which came across in the interview, and she is not afraid of an intellectual challenge. By the way, Jeanette uses mathematics at work "30,000 times a day, all the time."

Michael Fox graduated from the University of Michigan in 1993 with a mathematics major. He is currently in software design at Bell Northern Research/Northern Telecom. At the University of Michigan, there are two placement services, one through the school of Literature/Science/Arts and the other through Engineering. Michael used both services, but found the engineering one more useful. It seemed as though the companies interviewing there had more jobs. Most of the former were financial/statistics/consulting companies. He saw the engineering job descriptions from friends in the engineering school and decided that he could qualify for these jobs even though he did not have the formal qualifications listed.

Michael found it discouraging to do the job search with the engineers, but he did get some second interviews. He found that he had the technical background necessary for their jobs, plus he had a more well-rounded education. One real plus was his experience as a newspaper editor for the Greek system paper. He also had had a summer internship with a professor, during which he wrote a simulation of a mathematical model. This demonstrated to the employers that he could apply his background in Pascal and C, even though he did not have an extensive computer background.

Michael advises math majors to take writing courses. Don't limit yourself—apply for lots of different jobs. Keep an open mind and apply even when the job description says they want someone in computer science or electrical engineering. You can use this to your advantage. First you need to explain why you are qualified (even more qualified) for the job.

Julie Kallman graduated from Metropolitan State College of Denver as a mathematics major with a statistics/probability/applied mathematics emphasis. She is employed by the Siegel Company, which is an insurance and actuary consulting agency. Julie found this job through the coop center on campus. She worked for the Siegel Company part time while she was a student. In fact, she worked as an assistant in the department in which she is now working. For the full time job, the company needed specific skills, such as experience with Lotus 1-2-3 and Word Perfect, which Julie had. She also had experience with other office jobs, along with a business background and a degree in accounting. This on top of her mathematics and statistics background was appealing to her employer.

Julie's advice is to get experience even if it is not what you are looking for. It is most important to get into the work force, even if you intend to change jobs.

Chris Jepsen was a mathematics major at Grinnell College. At the time he graduated in 1993, he had almost completed a second major in economics. He is now working in an economics consulting company where one of his economics professors was once employed. His professor contacted the company, which led to an interview and the job. His qualifications include the ability to work with numbers in the form of data cleaning and data analysis. He also has the capacity to work with computers, although he did not have a formal background in computers. His mathematics major demonstrated analytic reasoning skills and strong logic skills to his employer. Chris's advice is to start early and consider many options. He did not know what he wanted, which put him at a definite disadvantage in the job market. Use the interview to help you focus on what you want to do. Chris plans to work one more year and then go to graduate school in economics.

Jill Stanley graduated from Lebanon Valley College as a mathematics major in 1993. She is working as a project scheduler at Star Expansion, a manufacturer of wall fasteners. Like many of the people interviewed, Jill had no idea what to do with her mathematics major. She only knew that she did not want to teach. A church connection who was an executive search person told her about manufacturing jobs and helped her make contacts. This was not an area that she would have thought of on her own.

The qualities needed for her job is a mathematics background, the ability to think things through from the beginning logically to the end, and lots of number crunching. She finds her job challenging since she had no production classes in college. Her advice is to look at manufacturing jobs for a career.

Ron Kirschman graduated a bit before the others. He finished at San Jose State University in December, 1989, with a mathematics degree. He is now a computer programmer at Lockheed. He found the job through the career center at San Jose State and an on-campus interview. However, he had worked at Lockheed before as an operator. He had quit this job to go back to school and get his degree, but the interviewer knew his work and he had very good references.

Ron's perspective is a bit different from the others since he has been employed longer. His advice is to stay in school and get a Master's degree. The job market is getting tighter, and he sees a large jump in salary and stability with a Master's degree over a Bachelor's. The competition is high, and the added degree makes a difference.

[illegible]

[illegible]

[illegible]

[illegible]

Paul Davis

So You Want to Work in Industry

The credentials for and the setting of a career in academic mathematics are fairly well understood: earn a doctorate in one of the mathematical sciences and work in a university department much like the one in which you were trained. University employers have two demands: Teach well and write good papers. The ratio shifts from institution to institution, but the fundamental ingredients are invariant.

The corresponding expectations of employers in business, industry, and government are less well known. Comments from more than forty nonacademic mathematicians describe an environment in which there are neither fixed expectations nor an assured niche for mathematicians.

Formal Training

Formal training in mathematics is much less a prerequisite for employment in business, industry, or government than in academia. And a good deal of what those trained as mathematicians do in industry might not qualify as mathematics by the standards of many universities. Literacy in some other field of science or engineering is often essential in industry, however.

Many people who do applied mathematics in industry have earned terminal degrees outside the mathematical sciences. For example, a consultant and head of a scientific computing group at a major pharmaceutical concern has a doctorate in chemistry. There are also individuals working as applied mathematicians who were trained in 'pure' mathematics. For example, a member of the research group for one of the big three auto manufacturers has a PhD in differential topology. Others have more conventional backgrounds in engineering, scientific computing, and applied mathematics.

With a touch of hyperbole, one mathematician with both academic and industrial experience addresses a different aspect of preparation for industry: "You need literacy in some field of science or engineering to get credibility or you need computer expertise. Unless it already had an institutionalized mathematics effort, [my company] wouldn't hire Leonhard Euler without a chemistry course."

Mathematicians who enter industry at the master's level may find themselves learning a specialty determined by their employer's needs, not by the details of their graduate training. That new specialty often has a large computational component. Training may come from formal in-house courses or as an inevitable response to job requirements. An applied mathematician in the aircraft industry suggests that the "most successful master's graduates are those who might as well have been PhD candidates. . . . The best master's hires grow to be almost PhD's."

Hiring Mathematicians

A harsh reality is the absence of a safe haven for mathematics outside academia. Few organizations in business, industry, or government have departments that must be staffed with mathematicians. And when there is a disciplinary bias because of a dominant product—chemicals, for example—the tilt is seldom toward mathematics. That is not to say there are not opportunities for applied mathematicians, especially with the increased use of mathematical modeling and computational simulation in industry. But the case has to be made.

Demonstrating relevance is a key to survival outside academia. On the desk of the physicist who is director of IBM Research, for example, is a paperweight with the message "Be Vital to IBM." A staff member in an automotive research laboratory says simply, "There is not a market niche for mathematicians."

A semiconductor manufacturer "doesn't usually hire mathematicians specifically. It hires to fill technical weaknesses—for example, device physics and modeling. The new hire may be a mathematician, or a physicist, or an electrical engineer. We know which universities produce the right people (at the doctoral level)."

The teamwork, communication skills, and breadth of scientific interest that are valued highly in industry contrast sharply with the characteristics of the single academic investigator working at the frontier of a subdiscipline. Indeed, generalists can often function more productively in industry than narrower specialists, particularly when a small group is called on to serve a diverse clientele. Critical judgment and problem-solving ability are always essential.

When there is a conscious decision to hire a mathematician, the desired characteristics might include those identified by one mathematician from a pharmaceutical concern: "a broad background and interest in a variety of mathematical areas, computation, and science in general." The other criteria used by this mathematician include the ability "to take a problem out of the blue" and the promise of "day-by-day professional development motivated by intellectual curiosity."

In the same vein, the technical center of a diversified metals manufacturer looks for "general problem-solving skills" and "the ability to communicate across disciplines." Whether among members of multidisciplinary teams or in the consulting environment in which many industrial mathematicians work, that ability to communicate across disciplines has a multitude of professional and interpersonal dimensions: listening skillfully, translating from another discipline into mathematics, maintaining visibility, integrating personal and professional relationships, and even simple salesmanship.

Attributes of Success

Much to the surprise of academic mathematicians, their industrial colleagues seldom list knowledge of specific subject matter when asked about the attributes of successful industrial applied mathematicians. Technical competence and (usually) knowledge of computation are taken for granted. The defining criteria are more cultural than purely intellectual.

In the aircraft industry, it may take six months to two years before new applied mathematicians can really make a contribution. It takes time to "teach them how to work in a group, to learn the corporate culture, to learn the nature of the company, and to master the nature of the problems and finding good solutions."

A prospective industrial applied mathematician must "show curiosity and the ability to penetrate." According to one industrial mathematician, "The key is an open mind and flexibility." Such individuals need to "know about the politics of work in industry (versus the academic environment in which they were trained), and they need to have more realistic expectations." They also need "taste for good methods and for good problems. We are all problem solvers in industry, whether we are mathematicians or marketers."

Communication and Interpersonal Skills

As one group of applied mathematicians quickly concludes, "Communication skills are key." According to another mathematician, the requirements extend beyond technical skills: "You need visibility for success. You must show others how and why your ideas work."

One member of this group observes that, "A mathematician needs communication skills to interact with chemists, physicists, and engineers of various stripes. Some cross-training helps you to get involved in problems at a much earlier stage. The cross-training that's important is not in a particular discipline. It is in the ability to approach a problem with an open mind, learning to translate from other disciplines into yours."

As suggested earlier, successful industrial work requires the ability to integrate personal and professional relationships. A consulting mathematician from a chemical company first learned at a cocktail party of a colleague's problem involving an important product for the analysis of blood chemistry. The friendship with that biochemist led to sharing the problem and then to the professional interactions of model building, analysis, and interpretation.

Another mathematician specializing in computation summarizes the importance of these personal and professional networks: "A lot of intelligence about the applications must go into software. Being in on the ground floor is essential to success. The more you see of a model, the more insight you have and the more political advantage you have."

The financial facts of life can make salesmanship essential. To find new funding in the face of budget cuts, one group at a government contractor now "must make friends and let them know our capabilities."

The importance of listening extends beyond the consulting environment. One applied mathematician describes a group of colleagues in a pure research organization "who were insulated from the real requirements of [our industry]. They looked at real problems with disdain. They preferred model problems, and they didn't know what the [real] business needs were. [People in the field] would ask 'What do I get for it? I can't use toy codes.' " This group's research charter lasted only as long as management was willing to protect it.

Clearly, industrial employers seldom value mathematicians for their own sake. Relevance counts, and mathematicians are valued only so long as they are relevant to the corporate mission.

Industrial Problems: Sources and Solutions

Solutions to good problems are the coin of the realm in the land of academic mathematics. Solving that first problem buys the doctorate. Later coins pay for promotion and tenure. Of course, the value of the coins can fluctuate with the fashions of the times. Results on catastrophe theory have depreciated like fins on a Chevrolet, while shape optimization holds its value like a Birkenstock sandal.

Although tastes change, the basic pattern varies little throughout academia. Good problems usually arise through the natural evolution of academic scholarship. Good solutions are presented at seminars and meetings before they are ultimately published. The value of a result is a subjective average of its elegance and novelty, and perhaps of the standing of the journal in which it is published.

The sources of industrial problems and the ways of valuing their solutions differ dramatically from the academic pattern. Conversations with a number of industrial mathematicians demonstrate that those differences sharply distinguish industrial mathematicians from their academic colleagues.

Sources of Problems

Many problems are posed to industrial mathematicians by colleagues in other disciplines, who may not yet understand the real problems they face. Good problems need not be elegant, new, or well posed, just necessary to corporate welfare.

Industrial problems, unlike those of academic scholarship, are seldom selected by natural evolution. For example, two applied mathematicians (one originally trained as a chemist) at a pharmaceutical manufacturer include in their suite of problems the development of a model of tumor heterogeneity. That problem was posed by the head of their laboratory.

Many mathematicians work either explicitly or implicitly in a consulting environment, which can provide a natural flow of problems. Because their clients "may not yet know the real problem, small questions can grow into big problems," as one industrial mathematician explains.

An applied mathematician working with a major computer graphics manufacturer observes that the problems "don't even have to be interesting—just necessary. If a group has hit the wall and their code release is next week, it's a good feeling when they make their deadline because you helped. You can go back to your other work with a sense of satisfaction." This same mathematician observes, "Every week I ask myself `Is my job secure from what I'm doing? Am I relevant? Am I known by others?' "

Finding the Real Problem

The key to a good solution is identifying the problem that really needs to be solved. In a very few cases, simply pushing back the frontier, either for internal dissemination or for external publication, is enough.

In most cases, an acceptable solution is a new piece that fits nicely into a larger puzzle that a multidisciplinary team is working to solve. Relevance and the quality of the fit can determine the value of the solution. Good solutions answer the question that really should have been asked, and they often are the consequence of deep involvement in problem formulation.

Communicating the solution to the user is important, especially since those users are usually not mathematicians. External publication, with a few exceptions, is much less highly valued than in academia and may even be restricted by corporate policy. Occasionally, a simple one-way transfer of information is adequate. But for one applied mathematician in the aircraft industry, who wants mathematics seen as an essential component of the company's success, that is not enough: "Simply saying, 'Here's the solution' sets the customer and the mathematician apart. It doesn't build a team. It doesn't contribute to having mathematics viewed as a key discipline like structural mechanics."

A basic requirement of a good solution is an understanding of the problem itself. At a major chemical manufacturer, that may mean "getting on the wavelength of a physical polymer chemist." For an applied mathematician with a large government contractor, understanding the problem requires "getting involved in the problem at a much earlier stage in order to capture its salient features."

An experienced consultant at a photographic products manufacturer observes that an applied mathematician "must hear the question that's really being asked. You must lead clients to see the real problems, not just dump a quick answer to the first question they ask." Others warn, "Be prepared to ask questions. [Ask the client,] 'What do you really want?' " "You must speak to engineers in a variety of disciplines and understand what their problems really are."

Using Solutions

In crafting a solution, mathematicians cannot be insulated from the competitive requirements of their businesses. Mathematicians "can't look at real-world problems with disdain or prefer model problems or not know what their companies' business needs are." "Someone has got to pay the bills."

Finally, timeliness is essential. An incomplete but timely solution is much more useful than one that is complete but too late. "Eighty percent of the solution in twenty percent of the time is more valuable," one industrial mathematician explained.

Exploiting the Trivial and Training Others

What the academic may scornfully dismiss as trivial, the industrial mathematician may need to exploit fully. Trivial problems can be important because they allow demonstrations of success and because their solutions can be used to build bridges to more important problems. An applied mathematician in the computer industry says, "You need a Mickey Mouse project where you can quantify progress."

Assisting in the solution of easy problems also provides opportunities to train new users, and hence additional advocates, of mathematics. For one internal consultant in a diversified chemical manufacturer, a chemists' request for help with the numerical solution of a system of seven ordinary differential equations was the beginning of a productive relationship that led to more challenging mathematics and significant contributions to profitable products.

The easy response for the mathematician would have been: "That's trivial. Use one of the packages in the computer center." But that answer would have pushed the chemist back across the disciplinary divide.

An applied mathematician in a group that develops software tools for engineering colleagues strikes a middle ground. "We usually provide the tools rather than teach the mathematical details. But for users to know when they have a mathematical problem, they need either good intuition of their own or salesmanship from our group." A mathematician in the pharmaceutical industry makes a case for developing "math awareness" but acknowledges that "we will hand-hold when the value is there" (that value being the development of another advocate for mathematics).

The relationships with professionals from other disciplines cover a spectrum of communication, teaching, indoctrination, and selling. A key to building the case for mathematics and for training nonmathematicians is openness and a participatory style. An applied mathematician working in computer graphics argues, "You have to use their language. It helps to talk out loud. They can understand the process you are going through. They can see that abstraction and turning back to the original context of the problem can be useful. People like to listen to how you think. You can't be just a black box. They must understand your solution."

The statistical consulting model is not appropriate for most other forms of industrial mathematics, according to an applied mathematician in the pharmaceutical industry. A statistical group that does "over-the-wall problem solving" (taking on whatever problems are thrown its way) can find itself with "high-priced people computing means because they don't train their users." In other settings it may be appropriate to "train the users, both as advocates for the solutions itself and as advocates for mathematics in general."

The matter of training colleagues can go far beyond the simple interchange of technical information. In one government contract laboratory, for example, a computational implementation of a complex model of groundwater flow and contanimant behavior is the basis for policy decisions that will affect the safety of community water supplies many centuries into the future. The final results of this modeling will influence the decisions made by government officials who have no conception of the power or the limits of mathematical modeling. In this setting, applied mathematicians become advocates for their profession as much as technical specialists. Nothing can be dismissed as trivial.

Another dimension of training is technology transfer. The importance of abetting the effective integration of new ideas cannot be dismissed quickly. Many users of mathematics are alert to new ideas, but "the technology transfer can't be one-way. You want the user to become a champion of the mathematics that's produced. You need a joint commitment to making the solution work."

In short, a good solution of an industrial problem is one that the user believes and uses, and a good problem is one that is central to corporate needs. Traditional academic standards of elegance and novelty are irrelevant. Some good problems may be easier than others, but none of them are ever trivial.

Paul Davis

Teamwork—The Special Challenge of Industry

The Working Environment

A faculty colleague of mine used to eat dinner regularly at our house. As Lou stepped out the door on his way home, his parting line was always, "See you tomorrow at the plant." We invariably smiled at the image of ourselves in chalk-covered overalls, carrying lunch buckets filled with books and pencils, on our way to a paper factory with ivy-covered walls.

Of course, for many industrial mathematicians "the plant" is just that, and the product is quite tangible, be it steel or software or strategies for pollution control.

What is it like to work in industry, away from the familiar patterns and the ephemeral products of academia? How do the habits and practices of those working environments compare with academia? How are those mathematicians supported?

Working with Colleagues

The usual industrial working environment ranges from large groups similar in size to university mathematics departments to isolated individuals. In any case, much of the work is done jointly, sometimes in rather large teams. The challenge of teamwork continues throughout a career, in contrast with the personally directed research path a tenured faculty member can choose to follow.

One industrial mathematician explains that the dictates of teamwork "may mean you have to do what you don't want to do for awhile." Another says, "You must have tolerance for a range of abilities and the wisdom to navigate the demands of teamwork and a diversity of personalities."

As one female mathematician makes clear, gender can be a factor in collegial relations: "Its tough for women if they are not aggressive. They must make sure people know what they did. They can't be afraid to say, 'That's my idea."

One mathematician's prescription for success emphasizes the importance of teamwork and skill with people: "Learn how to work together in teams, have an openness of mind and people skills. Bring in customers, and understand what they want. But understand that neither you nor they can know everything."

Sources of Support

Broadly speaking, industrial practitioners of mathematics are supported in three ways. (The rare exception is the laboratory with a pure re-search charter only loosely related to corporate productivity.) They may be part of a staff whose mission is directly linked to the company's product, production cycle, or service. Examples would be a mathematician developing signal processing algorithms for a defense contractor or a statistician responsible for quality control in a manufacturing plant.

The other two modes of support hinge on consulting. Those who function as consultants may be funded either directly from the corporate operating budget, often called funding from overhead, or they may be supported by billing their time to sponsors inside or outside the organization. Regardless of the source of funding and of problems, most industrial mathematicians agree that "being in the middle of the action pays."

Much like their academic colleagues who worry about the way the dean is treating their department, industrial mathematicians find that their support is often closely tied to the apparent value placed by upper management on the contributions of mathematics. In any case, mathematics is seldom the dominant technical discipline. At the corporate laboratories of a major, diversified chemical manufacturer, "Mathematics is always in the background. It is never in front with the physical problem. It is never in the limelight."

Breadth versus Depth

In the extreme case in which management practices intellectual apartheid, favoring one or two disciplines above all others, mathematicians are best hidden under the cloak of some other discipline. In happier situations, continuing efforts to point out the concrete contributions of mathematics can coincide with competitive forces to strengthen the support for mathematics. One major chemical company, for example, is expanding its group of mathematical consultants because of a competitive analysis showing that it can no longer afford unguided, Edisonian build-and-bust experiments. More intelligent modeling must inform its experimentation.

Although industrial employers do rely on narrow expertise, they often want breadth as well. In the pharmaceutical industry(and certainly elsewhere), "You do need years of experience to develop your craft," but practitioners also need breadth. "Industry wants breadth but relies heavily on narrow expertise as well," says one industrial mathematician.

The interdisciplinary work that is so common in industry also demands balance between breadth of knowledge and depth of knowledge. The latter alone is often the measure of mastery in an academic setting.

At one major corporate research laboratory, "The range of disciplines is so broad it doesn't matter what you know. Can you talk to others?" A mathematician who is an internal consultant to a petroleum company says, "I serve as a consultant. I can't specialize."

The size of the group with which the individual associates may determine the relative needs for depth and breadth. Larger groups of mathematicians can usually support a greater number of narrow specialists than smaller groups.

Cultural Barriers

Given the tendency of the corporate culture to favor certain disciplines (typically an engineering discipline) over mathematics, the introduction of mathematical approaches can be quite difficult. Moreover, a kind of glass ceiling in the management structure may allow those trained in one or two anointed disciplines to move into leadership roles while holding back mathematicians. Questions about the favoring of other disciplines over mathematics elicit explanations like "Nothing replaces the physical background."

Beyond those labeled as mathematicians, there is a larger community of users of mathematics and developers of computational tools. They are potential employers of mathematicians, and their work could be advanced by collaboration with well-educated interdisciplinary applied mathematicians.

Among such users, there may also be significant cultural barriers to the introduction of individuals who are trained primarily in mathematics. For example, engineers at a prominent defense contractor tell stories of lost competitive bids and design disasters that cry out for simple analyses and simulation. The corporate culture, however, is not ready for mathematics. Facing the strains of the end of the cold war, management has little interest in gambling on an unproven (and perhaps threatening) discipline.

From a different perspective, the vice president for research and development at a medium-sized manufacturer of precision optical instruments has a different reason for failing to make full use of mathematics: "We barely use arithmetic in our own quality studies. The central issue is culture—corporate and on the manufacturing floor—not mathematics."

Comparing Academia and Industry

A cultural gap also separates academic and nonacademic mathematicians. A day at the plant for an industrial mathematician is clearly much different from a day in the halls of the academy, and many industrial mathematicians think that academics don't appreciate the difference.

One industrial mathematician says, "My adviser and faculty treated me like I was lost when I decided to go into industry." An experienced independent consultant argues, "We may need to nurture attitude changes among ourselves that produce a comprehensive acceptance of a wide range of professional needs, not just those of the academic research mathematician."

For the industrial mathematician, interactions with both concepts and colleagues from a variety of disciplines are the norm. Such demands challenge industrial mathematicians to walk a tightrope between narrow expertise, the basis of academic research, and broader but necessarily more superficial knowledge. Much of the work of these mathematicians is determined by corporate policy, not individual research interests.

Such experiences, far removed from those of academia, are clearly the building blocks of satisfying, productive careers for many talented mathematicians. Perhaps those payoffs in productivity and personal satisfaction are just as deserving of the label "good mathematics" as anything that is published.

Suggestions for Students

What can students do to improve their chances of finding a good position in business, industry, or government? The suggestions that follow are strictly my own, but I believe you would hear similar messages from many others with experience outside of academia.

These suggestions are aimed primarily at graduate students because that population was studied by SIAM's Mathematics in Industry project, but undergraduates may find them useful as well. Many employers regard a bachelor's degree in mathematics as a sign of quantitative reasoning skill rather than a certification of specialized knowledge. They are prepared to train bachelor's hires to their organizations' specific needs. Remembering that difference, undergraduates should be able to extrapolate to their situation many of the suggestions given here.

Ask industrial managers what they seek in a good employee and you will hear answers like: "Flexibility, willingness to learn new areas of science." "A letter T — depth in an area of specialization (and) ... a broad understanding of the technical and business aspects of the company." "Team working and communication skills." "An interest in solving others' problems."

More formally, the primary demands of industrial employers include:

- the ability to work in teams,
- both communication (reading, writing, speaking, and listening) and interpersonal skills,
- breadth of scientific interest coupled with flexibility and an attitude of interest in the organization's work,
- the potential for continued professional growth,
- a broad technical background within mathematics as well as one other discipline, and,
- depth in an area of mathematics.

Good preparation for an industrial career should include experience in industry through internships or the like, plenty of practical problem solving, and greater emphasis on interdisciplinary studies.

You must have a genuine interest in working in industry. That interest will show in many subtle ways, and it will drive your choice of educational options, to the extent you have appropriate choices. Treat work in industry as a poor second to the university position you can't find, and you will have a tough time finding a job. And a tougher time being happy in it.

The mathematical requirements are relatively easy. Write a good thesis on a good problem. Use elective courses to acquire some mathematical breadth outside your specialty. Remember that your powers of abstraction, generalization, and logical reasoning distinguish you from an engineer, for example, with a strong mathematical background.

Be prepared to use those intellectual powers in settings that are different from academic mathematics. You need to translate problems and solutions back and forth from other disciplines.Those who pose problems to you and those who use your solutions will seldom be mathematicians. To build problem solving and communication skills, pursue seminar and problem solving opportunities in whatever form they present themselves.

Computational literacy seems essential. The individual who can't learn a new computer language, master a new operating system, or otherwise move easily through a computational environment is limited. As one industrial mathematician put it,"No computational skills? Then you are locked out." A computing course is probably useful, but actual experience in computational problem solving is better. That could be acquired through a summer job, thesis work, an internship, a clinic, or the like.

A good start toward interdisciplinary experience is a coherent series of courses in another department. Ideally, those will lead to evidence of real involvement with the vocabulary and with the problems and patterns of thought of the other discipline. Don't shy away from design, clinic, or laboratory courses if your background can accommodate them. A collection of watered down mathematics courses taught in another department is less useful.

Developing close relationships with faculty and fellow graduate students in another department can build contacts that are useful on several fronts. You will acquire day to day experience communicating (listening is even more important than talking!) with professionals outside of mathematics. You may also gain access to that discipline's network outside the university, a network that can help you find a summer job, a co-op position, an internship, or other kinds of substantial industrial experience.

Real experience outside the university can help you assess the depth of your interest in a nonacademic career. It can also give you substantial problem solving experience that can catch the eye of a potential employer who sees it on your resume. Entering the nonacademic job market without some significant exposure to industrial work can make your job search a good deal harder.

Experience in industry can also build interdisciplinary credentials and evidence of your ability to communicate across disciplinary boundaries. Finding this experience doesn't depend on your department's curriculum; look for a summer job or for one of the summer appointments offered by national laboratories. The university career office, the graduate school office, and contacts in engineering departments can all be useful sources of leads to temporary positions that will strengthen your background.

In searching for a short-term or a permanent position, you will not find many industrial employers waiting with open arms for mathematicians. You will find organizations in business, industry, and government who want to hire talented, flexible problem solvers with strong teamwork and interpersonal skills coupled with an instinct for the core issues of the business. Structure your mathematics education in light of those expectations without neglecting strong mathematical training and you should be competitive.

Note: The opinions expressed in this article are entirely my own, not official findings of SIAM's Mathematics in Industry study. Nonetheless, without passing to them any blame for errors in judgment or fact, I must acknowledge the insights and contributions of Avner Friedman, SIAM President, Jim Crowley, SIAM Executive Director, Bill Kolata, SIAM Technical Director, I. E.Block, retired SIAM Managing Director, and my colleagues on the MII Steering Committee, Bob Burridge, Peter Castro, Rosemary Chang, James Donaldson, Greg Forest, Bob Kohn, David Levermore, Sam Marin, Jim McKenna, Joyce McLaughlin, Bart Ng, Bob O'Malley, Jim Phillips, Rich Sincovec, and Lynn Wilson.

Paul Davis is a professor of mathematics at Worcester Polytechnic Institute.

Degrees Earned by Mathematics Majors

BA/BS Only

Cherryl Beeman
P. Darcy Barnett
Tiffany H. Brennan
Jack Cassidy
Shane Chalke
Joe Corrigan
H. Michael Covert
Sarah Cullen
Janet DenBleyker
Anitra Duckett
David Farmer
Janice A. Kulle
Tammy M. Lofton
Valerie Lopez
Stephen W. Louis
Carol J. Mandell
Carla D. Martin
Jennifer McLean
Deenanne Kay Myers
Diane R. Purcell
Laura Readdy
Brian D. Repp
Christine Rutch
Jeffrey A. Schneider
Deborah A. Southan
Robert L. Stewart
Mary Kay Esworthy Stiles
Kathy Haas Stukus
Lisa M. Sullivan
Elizabeth Sweet
Julie S. Tarwater
Mark Terry
Benjamin Weiss
Sarah Adams Yeary
Yvonne Zhou
Yaromyr Andrew Zinkewych

MA/MS Mathematics

Marion S. Ballard
Carl M. Beaird
Ron Bousquet
Margaret L. Brandeau
Judith R. Brown
Judith E. Chapman
James L. Cooley
Mark Derwin
M. Scott Elliott
Douglas A. Gray
Mary Hesselgrave
Patricia H. Jones
Kay Strain King
Loren Mernoff Lewin
Christopher M. Luczynski
Michael J. Murray
Joan Peters Ogden
Fred L. Preston
Mark A. Reynolds
William C. Schwartz
Frederick C. Taverner
Sue E. Waldman
Rodney B. Wallace
Gary L. Welz
Dan C. White

PhD Mathematics

Betsy Bennett
Yves Chiricota
Samson Cheung
Helaman Ferguson
Richard D. Fuhr
Holly Gaff
Ruth Gonzalez
Charles R. Hadlock
Charles Hagwood
John F. Hamilton
William D. Hammers
Carl M. Harris
Richard D. Jarvinen
Clare Johnson
J. Arthur Jones
Marcia P. Kastner
Maria Klawe
Michael H. Kutner
Peter A. Lindstrom
Patrick Dale McCray
Juan C. Meza
Harlan D. Mills
Michael G. Monticino
S. Brent Morris
William D. Murphy
Bruce A. Powell
Peter Rosenthal
David S. Ross
Bonita V. Saunders
Joel Schneider
Michael D. Weiss
Ana Witt
Larry Wos
Paul Zorn

Other Advanced Degrees

Jeffrey Z. Anderson	MS Industrial Engineering
Marion S. Ballard	MBA Real Estate
Carl M. Beaird	MS Physics
Jonathan G. Blattmachr	JD
Margaret L. Brandeau	PhD Engineering-Economic Systems
Denise Cammarata	MS Electrical Engineering
Cora L. Carmody	PhD Electrical Engineering (in process)
Mary K. Cooney	MS Secondary Education
Jeff Cooper	MS Telecommunications
Allison DeLong	MS Environmental Systems (Mathematical Modeling)
J. Wanzer Drane	PhD Biometry
Paula Duckett	MA Education
Thomas E. Dunham	PhD Metallurgy
Yawa Duse-Anthony	MS Industrial Engineering
Rol Fessenden	MA Geology
Frederick L. Frostic	MS Engineering
Holly Gaff	PhD Mathematics/Ecology
Cynthia C. Haddock	PhD Medical Care Organization and Administration
Mary Hesselgrave	PhD Formal Semantics
Harold Jacobs	MA Liberal Studies
Sherra E. Kerns	PhD Physics
Gary Kim	MS Business Administration
Greg King	MA Mathematics Education
Kay Strain King	Dip. Ed. Mathematics, Biology and Music
Karen M. Kneisel	MBA
Kimberly A. Kuntz	MS Information Science
Donna F. Lawson	MA Economics

Christopher M. Legault	PhD Marine Biology and Fisheries
Juan C. Meza	MS Electrical Engineering
Edward H. Preston	PhD Regional Planning
Fred L. Preston	PhD Industrial and Operations Engineering
Nancy K. Roderer	MLS Library and Information Science
Michael Rosenfeld	ME Mechanical Engineering
Peter Rosenthal	JD
Edward Rothstein	MA English Literature; PhD Committee on Social Thought
Julie Scarpa	MA Human Development
Joel Schneider	M. Ed. Mathematics Education
Mitchell Stabbe	JD
Arthur P. Staddon	MD
Debra W. Taufen	MM Finance and Accounting
Martha Weeks	MBA
Faedra Lazar Weiss	Master of Arts in Hebrew Letters
Michael D. Weiss	MA Economics
Sandra L. Winkler	MSE Industrial and Systems Engineering

Statistics

Susan J. Devlin
Cynthia C. Haddock
Jerry W. Highfill
Denise Ann Johnston
Michael H. Kutner
Mike Lieber
Marla Prenger
Jill L. Toll
Lisa M. Tonder
Richard M. White

Computer Science

Cora L. Carmody
Yves Chiricota
Marshall B. Garrison
Mary Hesselgrave
S. Brent Morris
Michael J. Murray
Edward H. Preston
Paul Willoughby

Steven G. Buyske*

Famous Nonmathematicians†

We often tell our students that there are many things besides teaching and actuarial work that they can do with a degree in mathematics, but I don't think they believe us. Over the years I've put together a list of well-known people who were math majors (or some equivalent in other countries and times), although not all of them completed their degrees. It's the most popular thing I've ever had on my office door. When I began this list, it had mostly contemporary Americans, and I called it "People who majored in math." Some of my students added their own names to their copies and posted them on their dorm doors.

I'd be delighted to hear of any additional names.

THE PUBLIC REALM

Ralph Abernathy, civil rights leader and Martin Luther King's closest aide.

Corazon Aquino, former President of the Philippines. She was a math minor.

Harry Blackmun, Associate Justice of the US Supreme Court, AB *summa cum laude* in mathematics at Harvard.

David Dinkins, Mayor of New York, BA in mathematics from Howard.

Alberto Fujimori, President of Peru, MS in mathematics from the University of Wisconsin-Milwaukee.

Ira Glasser, Executive Director of the American Civil Liberties Union, both a BS and an MA.

Lee Hsien Loong, Deputy Prime Minister of Singapore, a Bachelor's from Cambridge.

Florence Nightingale, pioneer in professional nursing. She was the first person in the English-speaking world to apply statistics to public health. She was also a pioneer in the graphic representation of statistics; the pie-chart was her inven-

*I'd like to thank my colleagues and the many people on USENET who have given me names and leads.

†Reprinted from the *American Mathematical Monthly*, Nov. 1993, pp. 845–847.

tion, for example. Not really a math major, she was privately educated, but pursued mathematics far beyond contemporary standards for women.

Paul Painlevé, President of France in the early 20th century, and one of the first passengers of the Wright Brothers. A ringer: he had a distinguished mathematical career.

Carl T. Rowan, columnist for the *Washington Post.*

Laurence H. Tribe, Professor at Harvard Law School, often regarded as one of the great contemporary authorities on Constitutional Law. An AB *summa cum laude* in mathematics from Harvard.

Leon Trotsky, revolutionary. He began to study Pure mathematics at Odessa in 1897, but imprisonment and exile in Siberia seem to have ended his mathematical efforts.

Eamon de Valera, long-time Prime Minister and then President of the Republic of Ireland. A ringer: he was a mathematics professor before Irish independence.

MUSIC

Ernst Ansermet, founder and conductor of the Orchestre de la Suisse Romande.

Pierre Boulez, Modernist composer and conductor.

Clifford Brown, Fifties jazz trumpeter.

Art Garfunkel, folk-rock singer. MA in mathematics from Columbia in 1967. Worked on a PhD at Columbia, but chose to pursue his musical career instead.

Phillip Glass, composer, a Bachelor's from the University of Chicago.

Carole King, Sixties songwriter, and later a singer-songwriter. She dropped out after one year of college to pursue her music career.

Tom Lehrer, songwriter-parodist. PhD student in mathematics at Harvard.

Lawrence Leighton Smith, conductor and pianist.

THE OTHER ARTS

Lewis Carroll, author of *Alice in Wonderland, Through the Looking Glass,* and other works. A ringer: he was a logician under his real name, Charles Lutwidge Dodgson.

Heloise (Poncé Cruse Evans), of *Hints from Heloise.* She minored in math.

Larry Niven, science fiction writer, winner of the Nebula and Hugo awards.

Omar Khayyam, author of *The Rubaiyat.* Another ringer: he published works on algebra and Euclid.

Alexander Solzhenitsyn, Nobel prize-winning novelist, a degree in mathematics and physics from the University of Rostov.

Bram Stoker, author of *Dracula,* took honors at Trinity University, Dublin.

Christopher Wren, the architect of St. Paul's Cathedral in London.